GLUCOCORTICOID HORMONE

GLUCOCORTICOID HORMONE
Mechanisms of Action

Edited by
Yukiya Sakamoto
and
Fumihide Isohashi

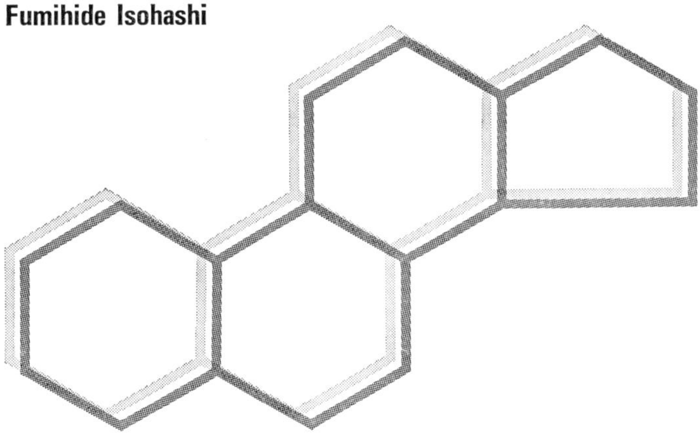

With 53 Figures and 18 Tables

JAPAN SCIENTIFIC SOCIETIES PRESS Tokyo
SPRINGER-VERLAG Berlin Heidelberg New York Tokyo

Supported in part by the Ministry of Education, Science and Culture under Grant-in-Aid for Publication of Scientific Research Result.

Published jointly by
Japan Scientific Societies Press Tokyo
ISBN 4-7622-9470-5
and
Springer-Verlag Berlin Heidelberg New York Tokyo
ISBN 3-540-15583-X Springer-Verlag Berlin Heidelberg New York Tokyo
ISBN 0-387-15583-X Springer-Verlag New York Heidelberg Berlin Tokyo

Sole distribution rights outside Japan granted to Springer-Verlag Berlin Heidelberg New York Tokyo

Printed in Japan

PREFACE

Almost 25 years have passed since Elwood V. Jensen and his colleagues at the University of Chicago first prepared a highly radioactive steroid hormone and demonstrated the longer retention of radioactive estrogen in the target tissue, suggesting the presence of a receptor in the tissue. Since then an intensification in the study of steroid hormone receptors and their molecular mechanisms of action has resulted in a great deal of information on how the steroid hormones stimulate cells by controlling the synthesis of particular proteins. The primary structure of the steroid hormone receptors and their subunits, and the modulators of receptor functions however have not previously been elucidated.

This book covers the latest advances in this and other aspects of research on glucocorticoid hormone receptors and the molecular mechanisms by which these hormones exert specific effects in target tissues. Later chapters are devoted to recent developments in other steroid receptor systems. Some of the data contained is appearing in print for the first time, and bibliographic information which will be helpful to the interested reader is also provided.

Contributors to the volume were invited speakers at the International Symposium on Mechanisms of Glucocorticoid Hormone Action held in Osaka, Japan in November 28, 1984. However, this is not a preceedings of the symposium, but has been published to further the understanding of the molecular mechanism of glucocorticoid hormone action which underlies the physiological process. It is our hope that it will serve to stimulate even greater strides in this area.

We are grateful to the Ministry of Education, Science and Culture of Japan for their Grant-in-Aid for publication of Scientific Research Results, and to the Kinki Branch of the Japanese Biochemical Society, the Alumni Association of the Department of Biochemistry of the Institute for Cancer Research, Osaka University Medical School, and the Yamanouchi Pharmaceutical Company for the financial support both to hold the symposium and to publish this.

November 28, 1985

Yukiya SAKAMOTO
Fumihide ISOHASHI

CONTENTS

1

Glucocorticoid Hormone: Mechanisms of Action, ed. Sakamoto, Y. and Isohashi, F., pp. 3–32, Japan Sci. Soc. Press, Tokyo/Springer-Verlag, Berlin (1986)

CONVERSION OF GLUCOCORTICOID RECEPTORS FROM A NONBINDING FORM TO A FORM THAT BINDS STEROID

EDWIN R. SANCHEZ, JOSEPH F. GRIPPO, PAUL R. HOUSLEY, and WILLIAM B. PRATT

Department of Pharmacology, The University of Michigan Medical School, Ann Arbor, Michigan 48109-0010, U.S.A.

As the binding of steroids to their receptors is a necessary event in the pathway by which they regulate gene expression and as the ligand binding event can be readily studied under cell-free conditions, our laboratory has focused on defining those factors that determine the ligand-binding state of the glucocorticoid receptor. Two types of receptor processing appear to be necessary for maintaining the steroid-binding conformation of glucocorticoid receptors in cell-free preparations. Reduction of receptor sulfur is absolutely required for steroid binding and receptor phosphorylation may be required, either directly by determining a steroid-binding conformation or indirectly by preventing disulfide bond formation. It is our purpose in this paper to first review some of our recent work defining the mechanism of receptor reduction in cytosol preparations. Secondly, we will show that the glucocorticoid receptor is a phosphoprotein and show how we are using ^{32}P labeling procedures to identify the existence of potential non-steroid-binding components of the molybdate-stabilized, untransformed glucocorticoid-receptor complex.

Thirdly, we will review some observations that have led to a model suggesting how phosphate and sulfur moieties on the receptor may interact in maintaining a stable steroid-binding conformation. The model also provides a useful way of thinking about how molybdate may act to stabilize the glucocorticoid receptor.

It is important to note at the outset that we will use the term "activation" throughout this paper to describe the process whereby the receptor is converted from a nonbinding form to a form that binds steroids. This is different from the use of the term "activation" to describe the process whereby the steroid-bound receptor is converted to a form that binds to nuclei, DNA cellulose, *etc.* We will use the term "transformation" to describe this latter process. Other work from this laboratory dealing with the glucocorticoid receptor cycle in intact cells, receptor inactivation and activation in cell-free systems, the endogenous receptor stabilizing factor, and the effects of group 6A transition metal oxyanions has been recently reviewed elsewhere (*2*, *5*, *14*, *15*) and will not be reviewed in this paper.

I. ACTIVATION AND INACTIVATION OF GLUCOCORTICOID-BINDING CAPACITY IN INTACT CELLS

Several laboratories have demonstrated that energy is required to maintain glucocorticoid-binding capacity in intact cells. Munck and Brink-Johnsen (*22*) were the first to show this when they demonstrated that specific glucocorticoid binding in intact rat thymic lymphocytes is reduced by deprivation of oxygen and glucose and rapidly returns to normal levels when aerobic conditions are restored. These alterations in binding are accompanied by a similar loss and reappearance of ATP. The thymocytes regain specific binding even in the presence of cycloheximide (*1*), suggesting that new protein synthesis is not required for receptor reactivation. On the basis of these observations Munck *et al.* (*23*) proposed that the glucocorticoid receptor in thymocytes exists in two states and that energy (ATP) is required to convert the nonbinding form to the steroid-binding form. This proposal is supported by other observations made in intact cells. When mouse fibroblasts (*17*), rat thymocytes (*27*), or chick embryo retina cells (*3*) are exposed to 2,4-dinitrophenol, glucocorticoid binding is lost. When the inhibitor is removed, steroid binding is restored, even if protein synthesis is inhibited (*3*, *17*, *33*).

In order to determine if the glucocorticoid receptor can be inactivated

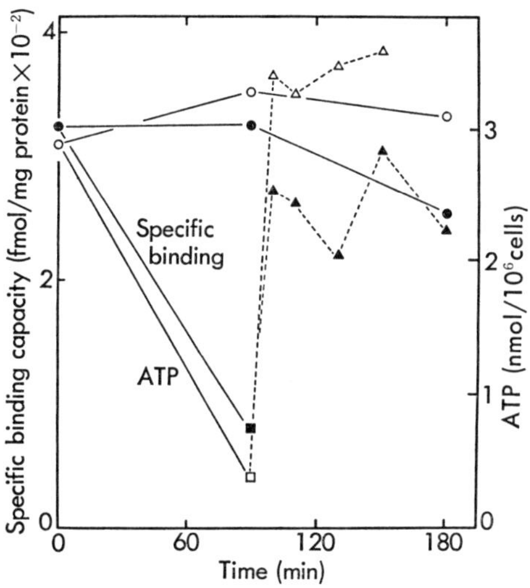

Fig. 1. Inactivation and reactivation of glucocorticoid-binding capacity in intact IM-9 cells. IM-9 cell suspensions were prepared in glucose-free medium and were incubated as follows: ●, plus glucose in air atmosphere; ■, minus glucose in nitrogen for 90 min; ▲, minus glucose in nitrogen for 90 min, then glucose was added and incubation was continued in air. At the indicated times, cell suspensions were harvested and assayed for specific glucocorticoid-binding capacity (solid symbols) and ATP (corresponding open symbols). The dashed lines connect the values obtained from cells that were preincubated under the energy-limiting conditions and then reactivated by addition of glucose and air. (From ref. *40*).

and reactivated in the absence of steroid, we examined the effect of metabolic restriction of glucocorticoid receptors in intact IM-9 human lymphoblasts that were not exposed to glucocorticoid prior to rupture (*40*). In the experiment shown in Fig. 1, IM-9 cells were incubated under normal nutritional conditions in air or under glucose-free conditions in an atmosphere of nitrogen. At various times, the cells were broken and the receptor activity of cytosol was determined by assaying the specific binding of [^{3}H]-triamcinolone acetonide. Under conditions of metabolic limitation, the steroid-binding activity and ATP content of the cells declines, and when glucose and oxygen are restored, both binding activity and ATP levels rapidly return to normal. This return of binding activity in IM-9 cells is not inhibited by cyclohexamide, suggesting that the changes represent a change in the binding state of the

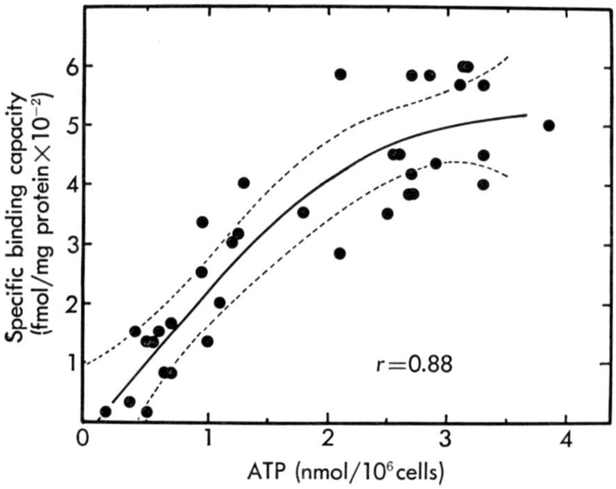

Fig. 2. Relationship between specific glucocorticoid-binding capacity and the amount of ATP in IM-9 human lymphocytes. The specific glucocorticoid-binding capacity assayed in low speed (27,000 × *g*) supernatants prepared from cells incubated with or without glucose and oxygen has been plotted against the ATP content assayed in the same cell suspensions. The solid line represents a least squares regression of the data and the dashed lines represent the 95% confidence limits for the regression. (From ref. *40*).

receptor and not in the number of receptor molecules in the cell (*40*). The clear correlation between glucocorticoid-binding capacity and ATP concentration in IM-9 cells is shown in Fig. 2. The fact that binding capacity can be altered raises the possibility that glucocorticoid receptors may be switched on and off in a rapid, reversible manner, and it is possible that a cell's capacity to respond to glucocorticoid could be modulated by altering the fraction of active receptors.

II. THE ENDOGENOUS RECEPTOR REDUCING SYSTEM

Rees and Bell (*27*) demonstrated in 1975 that glucocorticoid receptors in cytosol prepared from rat thymocytes must be reduced in order to bind steroids. Granberg and Ballard (*8*) subsequently proposed that the glucocorticoid-binding capacity of cytosol preparations reflected the endogenous sulfhydryl reducing activity of the system. Often, biochemists add reducing agents to cell-free systems and tend to ignore the endogenous processes by which various proteins are reduced. As the alterations in glucocorticoid-binding capacity that occur in intact cells could reflect redox changes in the receptor, in recent months we have spent considerable effort trying to define the endogenous system that reduces the receptor in rat liver cytosol (*9*, *10*).

The system we have used to assay for receptor activation is rat liver cytosol containing receptors that have been inactivated by extracting the cytosol with 10 % charcoal (*9*). The experiment of Table I shows the effect

TABLE I
Effect of Charcoal on Unbound and Steroid-bound Receptor

Conditions	Specific binding or Specific binding capacity: Before charcoal extraction (cpm/0.2 ml cytosol)	After charcoal extraction (cpm/0.2 ml cytosol)
Unbound cytosol	20,400	1,140
Steroid-bound cytosol	21,800	19,200

Rat liver cytosol was divided into two portions. One portion was extracted with 5 % charcoal for 30 min and then incubated with [^{3}H]-triamcinolone acetonide for 2.5 hr to determine specific binding capacity. Another portion of rat liver cytosol was incubated with [^{3}H]-triamcinolone acetonide for 2.5 hr in the presence or absence of excess dexamethasone and the steroid-bound cytosol was extracted with 5 % charcoal for 30 min. The specific binding in each sample was determined before and after charcoal extraction. In this experiment, bound steroid was separated from free steroid by passing each sample through short columns of Sephadex G-25 as we have previously described (*16*). All procedures were carried out at 0–4°C. (From ref. *9*).

TABLE II
Reactivation of Receptors in Charcoal-treated Liver Cytosol with Dithiothreitol or Boiled Liver Cytosol

Condition	Specific binding capacity (cpm/0.2 ml cytosol)
Untreated receptor preparation	32,600
Charcoal-treated receptor preparation	792
+Dithiothreitol	25,500
+Boiled cytosol	28,500

Rat liver cytosol was extracted with 10 % charcoal for 30 min at 0–4°C to inactivate glucocorticoid receptors and then incubated in the presence of 10 mM sodium molybdate for 3 hr at 20°C, either alone or with 10 mM dithiothreitol or with boiled liver cytosol. Following incubation of the mixtures at 20°C, the specific binding capacity of each sample was assayed at 0°C. (From ref. *9*).

of charcoal extraction on unbound and steroid-bound receptors in rat liver cytosol. As expected, charcoal extraction has little or no effect on the steroid-bound receptor, but extraction of cytosol with 5 or 10 % charcoal at 0°C inactivates unoccupied receptors (*9*). It is clear that charcoal extraction has in some way promoted oxidation of the receptor because, as shown in Table II, the binding capacity can be returned by adding dithiothreitol. The binding capacity can also be returned by incubating charcoal-extracted cytosol at 20°C with molybdate and boiled rat liver cytosol (Table II). Thus, rat liver cytosol contains an endogenous heat-stable reducing system that activates glucocorticoid receptors to the steroid-binding form.

In the experiment of Fig. 3, charcoal-extracted rat liver cytosol was passed through a column of Sephadex G-50 (in molybdate containing buffer) to separate the inactive receptors from small Mr components of cytosol

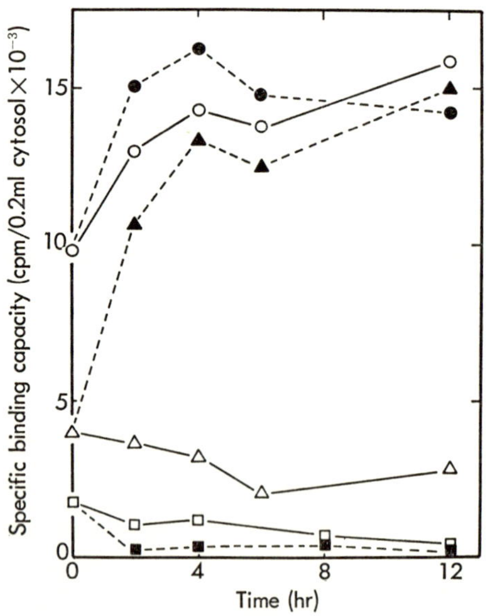

Fig. 3. Time course and temperature-dependence of receptor activation by dithiothreitol and boiled cytosol. Charcoal-treated, Sephadex G-50-filtered cytosol was incubated in the presence of 10 mM sodium molybdate for various times under the following conditions: control incubation in buffer alone at 20°C (■) or at 0°C (□); plus 10 mM dithiothreitol at 20°C (●) or at 0°C (○); plus boiled liver cytosol at 20°C (▲) or at 0°C (△). The specific binding capacity of untreated cytosol was 21,700 cpm/0.2 ml. (From ref. *9*).

(*e.g.*, glutathione, NADPH, thioredoxin) that are involved in protein reduction. As shown in the figure, reactivation of receptors in Sephadex G-50-filtered cytosol by dithiothreitol is temperature-independent, but reactivation by boiled cytosol is clearly temperature-dependent, suggesting that enzymatic activity is required for activation by the endogenous system. We have taken apart the receptor activating system in boiled cytosol and find that two heat-stable components are required for receptor activation.

To separate the potential reducing components in boiled cytosol from each other, boiled cytosol was chromatographed on a column (2.1 × 34 cm) of Sephadex G-25. The fractions containing the macromolecular material, as defined by the peak of absorbance at 280 nm, were combined, lyophilized to dryness, and reconstituted in sufficient buffer to yield the concentration of macromolecular components (G-25 macro) originally present in boiled cytosol after dilution in the receptor reactivation assay. The rest of the fractions which contain the micromolecular material (G-25 micro) were combined and treated in the same way. This procedure yields a heat-stable macromolecular preparation which contains most of the thioredoxin and no NADPH or glutathione, all of which were independently assayed in each fraction from

TABLE III
Reconstitution of the Endogenous Receptor Activating System

Addition	Specific binding capacity (cpm/0.2 ml cytosol)
Boiled cytosol	30,000
G-25 macro	2,490
G-25 micro	11,310
G-25 macro + G-25 micro	31,400
G-25 macro + 25 μM NADPH	23,900
G-25 macro + 2 mM glutathione	8,320
G-25 macro + 25 μM NADPH + 2 mM glutathione	26,700

Boiled rat liver cytosol was chromatographed through a column of Sephadex G-25. The fractions eluting with the macromolecular peak at A_{280} (G-25 macro) were combined, lyophilized, and redissolved in one-half the original volume of Hepes buffer. The remaining fractions containing all of the smaller heat-stable components of cytosol were combined, treated in a similar manner, and designated G-25 micro. Charcoal-extracted, Sephadex G-50-filtered cytosol was incubated with the indicated additions for 2 hr at 20°C in the presence of 10 mM sodium molybdate. Following the 20°C incubation, charcoal-inactivated, Sephadex-filtered cytosol alone bound 1,690 cpm, plus 10 mM dithiothreitol 29,600, plus 25 μM NADPH 4,240, plus 2 mM glutathione 6,710, plus NADPH and glutathione 9,200 cpm/0.2 ml cytosol. (From ref. *9*).

the Sephadex G-25 column. The micromolecular preparation contains all of the NADPH and glutathione, and a small amount of thioredoxin. As shown in Table III, the macromolecular fraction of boiled cytosol does not activate glucocorticoid receptors in charcoal-extracted, Sephadex G-50-filtered cytosol and the micromolecular fraction yields only a small amount of activation.

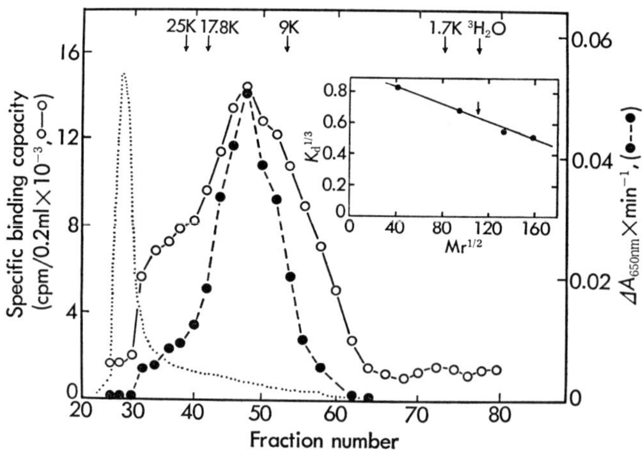

Fig. 4. Coelution of receptor activating activity and thioredoxin activity on Sephadex G-50 chromatography of boiled cytosol. The endogenous receptor activating factor in boiled rat liver cytosol was separated from NADPH by Sephadex G-25 chromatography and the Sephadex G-25 macromolecular peak was concentrated and chromatographed on a long Sephadex G-50 column. Aliquots of every other fraction from the Sephadex G-50 column were assayed for their ability to activate receptors in charcoal-inactivated, Sephadex G-50-filtered cytosol in the presence of 100 μM NADPH and 10 mM sodium molybdate (○). The inactivated cytosol alone bound 890 cpm, in the presence of 100 μM NADPH, 1,430 cpm, and in the presence of 10 mM dithiothreitol, 20,300 cpm/0.2 ml cytosol. Thioredoxin was assayed in every other fraction by the insulin disulfide reduction method of Holmgren (*11*) and the activity is presented as the rate of change in turbidity at A_{650} per aliquot (●). The Sephadex G-50 column was calibrated using α-chymotrypsin (Mr=25,000), myoglobin (Mr=17,800), lima bean trypsin inhibitor (Mr=9,000) and vitamin B_{12} (Mr=1,700) as molecular weight markers. Blue dextran and tritiated water were used to determine V_0 and V_i respectively. The inset shows the relationship between the distribution coefficient (K_d) of the marker proteins and their molecular weights. The arrow in the inset indicates the apparent Mr of the endogenous activating activity at 12,300. (From ref. *9*).

When the macromolecular and micromolecular components of boiled cytosol are combined, however, receptor activating activity is fully restored. Substantial receptor activation is achieved with the macromolecular fraction of boiled cytosol and NADPH. The macromolecular fraction and 2 mM glutathione (the concentration of glutathione assayed in boiled liver cytosol) do not activate beyond the minimal level obtained with glutathione alone, suggesting that NADPH is the important micromolecular reducing component that acts with the macromolecular factor to produce a coupled reducing system that activates the receptor.

On the basis of its physical behavior, it seemed likely that the macromolecular component of the reducing system could be thioredoxin, a heat-stable reducing protein that has been purified from rat liver and has an Mr of 12,000 (*24*). Accordingly, the Sephadex G-25 macromolecular material was rechromatographed on Sephadex G-50 and every other fraction was assayed both for thioredoxin content by the insulin disulfide reduction assay (*11*) and for its ability to reactivate receptors in Sephadex G-50-filtered cytosol in the presence of NADPH. As shown in Fig. 4, the two activities coelute from the column in a major peak (apparent Mr of about 12,300) with a shoulder that may represent thioredoxin dimer. Fractions 35 to 58 of the Sephadex G-50 column shown in Fig. 4 were pooled, lyophilized to dryness, and dissolved in 2 ml of distilled water. The concentrated buffer was removed by Sephadex G-25 chromatography and the desalted material was concentrated by lyophilization and stored in small aliquots at −20°C. We have found that repeated freezing and thawing is accompanied by loss of NADPH-dependent receptor activating activity. Thus, the preparation is frozen in small lots and thawed once. We have called this partially purified, NADPH-free, heat-stable material the endogenous receptor activating factor or EAF.

As shown in Table IV, this endogenous activating factor and purified *Escherichia coli* thioredoxin activate receptors in Sephadex G-50-filtered cytosol in a similar manner, with the activation by both compounds being entirely NADPH-dependent. Rat liver thioredoxin reductase (native molecular weight of 116,000) is present in the Sephadex G-50-filtered receptor preparation and it is well established that the rat liver enzyme reduces *E. coli* thioredoxin in the presence of NADPH (*20*). As the endogenous activating factor is a heat-stable protein (it is inactivated by papain) that coelutes with endogenous thioredoxin activity on Sephadex G-50 chromatography, as

TABLE IV
Activation of Receptors by NADPH and the Endogenous Activating Factor (EAF) or Purified *E. coli* Thioredoxin

Condition	Specific binding capacity (cpm/0.2 ml cytosol)
Charcoal-treated, Sephadex G-50-filtered cytosol	
alone	1,320
+Dithiothreitol	24,000
+NADPH	4,390
+EAF	1,000
+Thioredoxin	1,010
+EAF+NADPH	20,600
+Thioredoxin+NADPH	16,700

The endogenous, heat-stable glucocorticoid-receptor activating factor (EAF) was purified from boiled rat liver as described by Grippo *et al.* (*14*). Charcoal-inactivated liver cytosol was passed through a column of Sephadex G-50 to separate the glucocorticoid receptor from NADPH and thioredoxin. The inactive receptor was incubated for 2 hr at 20°C with the indicated additions and specific binding was assayed at 0°C. Additions were added at the following concentrations: dithiothreitol, 10 mM; NADPH, 100 μM; *E. coli* thioredoxin, 5 μM; and concentrated EAF at 40 % of the incubation volume. (From ref. *9*).

it activates receptors in an NADPH-dependent manner, and as it can be replaced in the activating system by purified *E. coli* thioredoxin, it is reasonable to conclude that the activating factor is thioredoxin. Since NADPH alone does not activate the Sephadex G-50-filtered receptor, the immediate donor of reducing equivalents must be thioredoxin.

In order to further test the proposal that the endogenous heat-stable, receptor-activating system is an NADPH-dependent reduction mediated by thioredoxin, we have examined the effects of antibodies to rat liver thioredoxin and thioredoxin reductase on the activation of glucocorticoid-binding capacity (*10*). Thioredoxin and thioredoxin reductase were purified to homogeneity from rat liver cytosol by the method of Luthman and Holmgren (*20*) and the purified preparations were used to raise specific antisera in rabbits. The effects of the antisera were examined in charcoal-extracted cytosol. Charcoal-extraction of cytosol removes NADPH but it does not remove thioredoxin. Thus, in contrast to cytosol that has been passed through a column of Sephadex G-50 to remove thioredoxin, receptors in charcoal-extracted cytosol can be reactivated by addition of NADPH. As shown in Fig. 5A, thioredoxin reductase antiserum prevents activation of receptors caused by incubating charcoal-treated cytosol with NADPH. Antiserum raised against purified rat liver thioredoxin also inhibits NADPH-mediated receptor activation, although it is not as effective as antiserum raised against the reductase. Figure 5B shows the immune-specific inhibition of receptor activation by boiled rat liver cytosol that is produced by antiserum to thioredoxin reductase. The thioredoxin reductase antiserum is less effective in preventing receptor reactivation caused by boiled cytosol (B) than activation caused by NADPH (A) because the boiled cytosol contains (in addition to NADPH and oxidized thioredoxin) some reduced thioredoxin which participates in direct receptor activation (*9*), thus increasing the specific binding capacity without the requirement for thioredoxin reductase activity. Thioredoxin antiserum is not effective at preventing receptor reactivation by boiled cytosol. This may be due to the fact that the addition of boiled cytosol (thioredoxin concentration

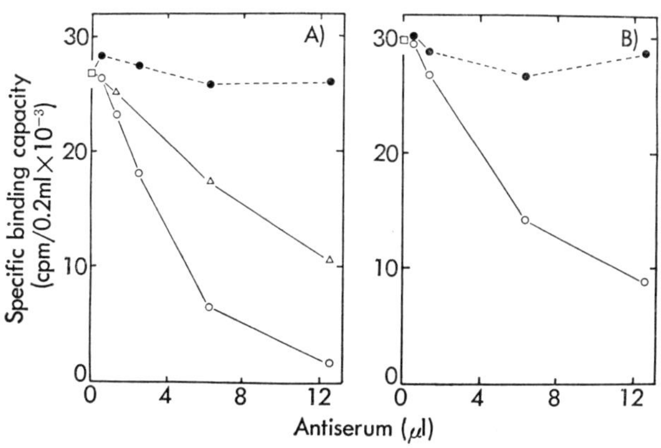

Fig. 5. Effect of thioredoxin reductase-specific antibodies on receptor reactivation in charcoal-treated ctyosol by NADPH and boiled liver cytosol. A 100,000 ×*g* supernatant of rat liver was extracted with 10 % charcoal for 30 min on ice. The charcoal-treated cytosol was mixed with sodium molybdate (10 mM) and incubated for 30 min at 20°C with 10 mM Hepes buffer (□), antiserum to thioredoxin reductase (○), antiserum to thioredoxin (△), or preimmune control serum (●). Following the preincubation period, NADPH (Panel A) at 100 μM or boiled liver cytosol (Panel B) at 25 % (v/v) of the final incubation volume was added and the incubations were continued for 2 hr at 20°C. At the end of the incubation, specific glucocorticoid-binding capacity was determined at 0–4°C. Charcoal-treated cytosol alone bound 1,440 cpm; plus NADPH (100 μM), 26,900 cpm; plus dithiothreitol (10 mM), 31,400 cpm; and plus boiled rat liver cytosol (25 % v/v), 30,000 cpm/0.2 ml cytosol. (From ref. *10*).

of ~4 μM *vs.* ~2 μM for charcoal-treated cytosol, ref. *9*) triples the thioredoxin content of the system.

Taken together, the data presented here demonstrate that the endogenous glucocorticoid-receptor activating factor is thioredoxin, and we would propose that the maintenance of specific glucocorticoid-binding capacity in the intact cell may require the continuous availability of reduced thioredoxin. There are only a couple of instances where the mechanism of thioredoxin action has been examined in detail. Thioredoxin was first discovered as the donor of reducing equivalents for the ribonucleotide reductase reaction (*36*). In this system thioredoxin acts *via* a ping pong mechanism in which it first forms a mixed disulfide with the reductase which then readily yields the reduced enzyme and oxidized thioredoxin. In this case, it has not been possible to demonstrate thioredoxin-enzyme complexes by physical means, such as by equilibrium dialysis (*35*). In contrast to the ribonucleotide reductase example, thioredoxin has been found to be a necessary subunit of the active T7 DNA polymerase molecule (*21*). The interaction between reduced thioredoxin and the 80K subunit of DNA polymerase is of high enough affinity to allow purification of the polymerase enzyme using antithioredoxin immunoadsorbent chromatography (*24*).

At this time, we do not know whether thioredoxin simply reduces a disulfide bond to yield a reduced, steroid-binding form of the glucocorticoid receptor or whether, as in the T7 DNA polymerase example, reduced thioredoxin must be bound to the receptor for it to be in a steroid-binding conformation. Another possibility is that thioredoxin acts either enzymatically or structurally upon a nonreceptor component that subsequently affects receptor function. If thioredoxin occupies a specific binding site on the receptor, then there must be a specific thioredoxin-binding domain. In this light, it is interesting to note that Schmidt *et al.* (*30*) have inactivated the glucocorticoid-binding capacity of rat liver cytosol by incubating it with polyvalent antiserum prepared against partially purified calf thymus DNA polymerase α (*24*), leading them to suggest that glucocorticoid receptors and DNA polymerase α may share antigenic determinants. Tanaka *et al.* (*34*) have prepared several monoclonal antibodies against human DNA polymerase α and have found that the monoclonal antibodies failed to reduce the dexamethasone-binding capacity of human KB cell cytosol and failed to precipitate the dexamethasone-receptor complex (*18*). The difference between the two observations could be explained if the polyclonal antiserum used by Schmidt *et al.* (*30*) contained antibodies that interacted with thioredoxin or with antigenic determinants in a thioredoxin-binding domain common to both DNA polymerase α and the glucocorticoid receptor.

III. LABELING OF THE GLUCOCORTICOID RECEPTOR AND A POTENTIAL RECEPTOR-ASSOCIATED PROTEIN WHEN L-CELLS ARE GROWN IN THE PRESENCE OF [^{32}P]-ORTHOPHOSPHATE

In 1983, we reported (*13*) that molybdate-stabilized cytosol prepared from mouse L-cells that had been incubated for 18 hr in the presence of [^{32}P]-orthophosphate contains two phosphoproteins (a 90–92K and a 98–100K protein) that elute from an affinity resin of deoxycorticosterone-derivatized agarose in a manner consistent with the predicted behavior of the glucocorticoid receptor. The assignment of these two phosphoproteins as the glucocorticoid receptor was based on three observations. First, the major 90–92K phospho-

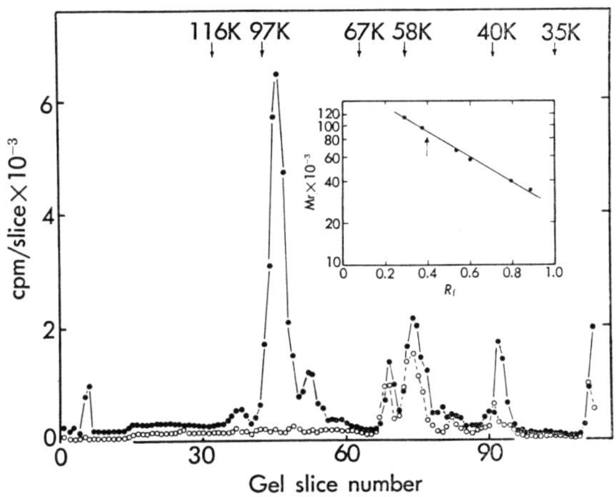

Fig. 6. SDS-polyacrylamide gel electrophoresis of [^{3}H]-dexamethasone mesylate-labeled L-cell cytosol. Cytosol was incubated at 0–4°C with 50 nM [^{3}H]-dexamethasone mesylate and either vehicle (●) or 50 μM unlabeled dexamethasone (○), and submitted to SDS-polyacrylamide gel electrophoresis. Sample lanes were cut into 1 mm slices, digested, and radioactivity was determined by liquid scintillation counting. The inset illustrates the standard curve used to obtain the Mr of the covalently labeled receptor using proteins of known Mr. (From ref. *13*).

protein and the minor 98–100K phosphoprotein were the same size as the L-cell glucocorticoid receptor identified by site-specific labeling with [^{3}H]-dexamethasone 21-mesylate (*13*). Dexamethasone 21-mesylate is a glucocorticoid derivative developed by Simons *et al.* (*31*) that contains a reactive C-21 methanesulfonyl group. Dexamethasone mesylate has been shown to covalently label the glucocorticoid receptor of HTC cells (*32*). As shown in Fig. 6, L-cell cytosol that was labeled with 50 nM [^{3}H]-dexamethasone mesylate contained two proteins that were labeled in a specific manner. When proteins were submitted to sodiumdodecylsulfate (SDS)-polyacrylamide gel electrophoresis, most of the specifically bound radioactivity migrated slightly faster

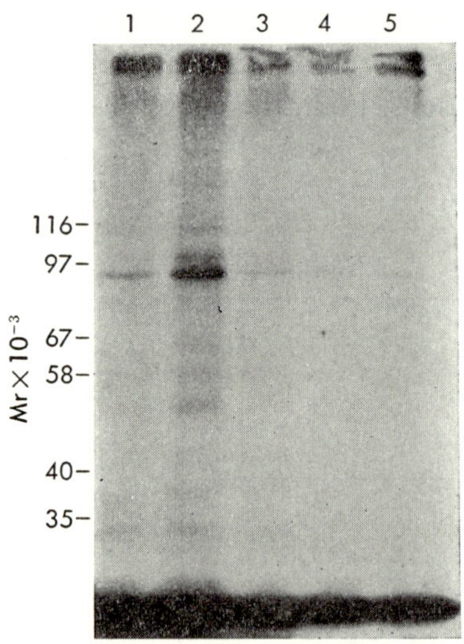

Fig. 7. Stereospecific elution of ^{32}P-labeled glucocorticoid receptor from deoxycorticosterone-agarose. ^{32}P-labeled L-cell cytosol was incubated with the affinity agarose to bind receptor and the column was eluted with the indicated isomer of cortisol. Following SDS-polyacrylamide gel electrophoresis, the [^{32}P]-phosphoproteins were visualized by autoradiography. Lane 1, elution with the physiologically inactive compound 11α-cortisol; lane 2, elution with the active glucocorticoid 11β-cortisol; lane 3, cytosol was preincubated with 50 nM triamcinolone acetonide and then incubated with the affinity column and eluted with 11β-cortisol; lane 4, 11α-cortisol elution of ^{32}P-labeled cytosol from glucocorticoid-resistant L-cells; lane 5, 11β-cortisol elution of ^{32}P-labeled cytosol from resistant L-cells. (From ref. *13*).

than the phosphorylase marker at 90–92K and a minor portion migrated slightly slower than the marker at 98–100K.

In the experiment shown in Fig. 7, L-cells in monolayer culture were grown in the presence of [^{32}P]-orthophosphate for 18 hr prior to harvest. The glucocorticoid-binding proteins in cytosol were purified by affinity chromatography on a column of deoxycorticosterone-agarose and submitted to SDS-polyacrylamide gel electrophoresis. In lane 1 the affinity column was eluted with 11α-cortisol which has no glucocorticoid activity and in lane 2 the active isomer 11β-cortisol was present. The presence of the active glucocorticoid results in increased elution of both a major phosphorylated species migrating at 90–92K and a less prominent species that migrates slower than the phosphorylase marker at 98–100K. In order to test if these phosphoproteins bind glucocorticoids in a high affinity manner, ^{32}P-labeled cytosol was preincubated with 50 nM triamcinolone acetonide in order to occupy the receptor sites prior to exposure to deoxycorticosterone-agarose. There is much less of the 90–92K species and none of the 98–100K species in this preparation (lane 3). These observations suggested that both phosphoproteins bound glucocorticoids in a stereospecific and high affinity manner and formed a second basis for proposing that they represented the glucocorticoid receptor. The third observation relating these two phosphoproteins to the glucocorticoid receptor is shown in lanes 4 and 5 of Fig. 7. In this case a cloned glucocorticoid-resistant subline of L929 cells with only 5 percent of the binding capacity of the parent cells was incubated with [^{32}P]-orthophosphate and the cytosol proteins were submitted to affinity chromatography. The resistant cells contain much less phosphoprotein migrating at 90–92K and no phosphoprotein migrating at 98–100K.

At the time this work was carried out, we did not know if the 98–100K phosphoprotein was the unaltered receptor and the 90–92K phosphoprotein was a cleavage product or if these two steroid-binding phosphoproteins were products of different genes. While we were proceeding with experiments designed to purify the 90–92K phosphoprotein, we noted that we were recovering more ^{32}P radioactivity in the 98–100K region of the eluate from the affinity column than we had in our original studies. We therefore eluted the affinity column with [^{3}H]-dexamethasone mesylate and found that the major species labeled by the affinity ligand in a manner that was prevented by coelution with competing non-radioactive dexamethasone migrated at 98–100K. Thus, we had a shift to more 98–100K labeling by both ^{32}P and [^{3}H]-dexamethasone mesylate, implying that less cleavage was occurring. Accordingly, we again examined the binding of [^{3}H]-dexamethasone mesylate in L-cell cytosol.

To isolate the receptor-bound [^{3}H]-dexamethasone mesylate from other components of cytosol we used the monoclonal anti-receptor antibody GR49. This antibody was prepared against the rat liver glucocorticoid receptor but it cross-reacts with the mouse receptor (*39*). [^{3}H]-Dexamethasone mesylate-bound receptors were incubated with the monoclonal antibody, adsorbed to protein-A-Sepharose, and resolved by SDS-polyacrylamide gel electrophoresis. We found that the only [^{3}H]-dexamethasone mesylate-bound protein brought down by the antibody migrated at 98–100K. The antibody also brought down a 90–92K Coomassie Blue-staining band that did not bind the affinity

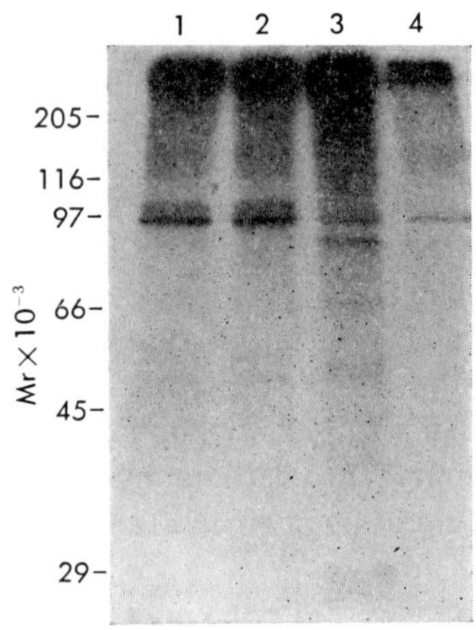

Fig. 8. Extraction of ^{32}P-labeled proteins from L-cell cytosol with monoclonal antibody and protein-A-Sepharose. Four 400 μl aliquots of ^{32}P-labeled L-cell cytosol were mixed with an equal volume of buffer and bound with 50 nM non-radioactive dexamethasone. Each aliquot was incubated for 12 hr on ice with monoclonal antibody, or with non-immune mouse IgG (each at 5 % of volume), or with no immunoglobulin. At the end of the incubation, each sample was bound to protein-A-Sepharose for 2 hr. The protein-A-Sepharose-bound material was washed as described in ref. *30*, and samples were resolved by electrophoresis on a 9 % SDS-polyacrylamide gel and autoradiography. Lane 1, ammonium sulfate purified GR49 monoclonal antibody; lane 2, the same antibody purified on protein-A-Sepharose; lane 3, non-immune mouse IgG; lane 4, no immunoglobulin addition. (From ref. *16*).

ligand. As shown in Fig. 8, when cytosol from ^{32}P-labeled L-cells is reacted with the monoclonal antibody we recover a ^{32}P-labeled band at 98–100K that is not seen with non-immune mouse immunoglobulin G (IgG) and a marked increase in ^{32}P-labeling in the 90–92K region over that seen in the non-immune control lane.

The experiment of Fig. 9 was performed to show that the 98–100K protein labeled with [^{3}H]-dexamethasone mesylate and the 90–92K, ^{32}P-labeled protein can clearly be resolved from each other. L-cell cytosol labeled

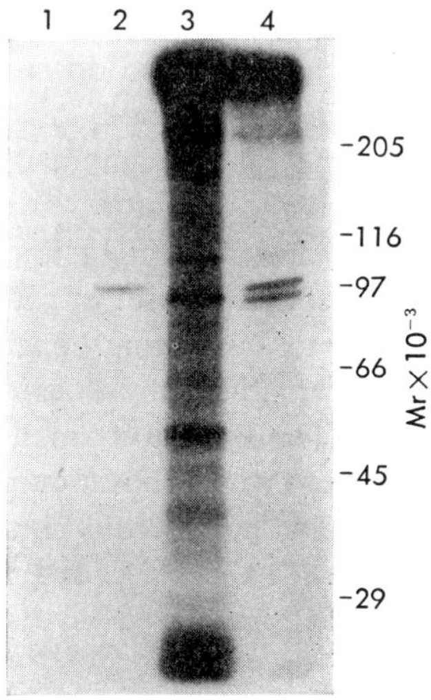

Fig. 9. Comparison of [^{3}H]-dexamethasone mesylate-labeled receptor with ^{32}P-labeled L-cell cytosol proteins after reaction with monoclonal antibody. L-cell cytosol was incubated with 50 nM [^{3}H]-dexamethasone mesylate in the presence or absence of 1000-fold excess of non-radioactive dexamethasone, and 300 μl of each incubation was further incubated with monoclonal antibody and protein-A-Sepharose as described in ref. *30*. Aliquots (100 μl) of ^{32}P-labeled L-cell cytosol were incubated with monoclonal antibody and protein-A-Sepharose in the same manner. Samples applied to each lane of a 7 % SDS-polyacrylamide gel were: lane 1, 300 μl of cytosol incubated with [^{3}H]-dexamethasone mesylate in the presence of excess non-radioactive dexamethasone; lane 2, 300 μl of cytosol incubated with [^{3}H]-dexamethasone mesylate alone; lane 3, 100 μl of ^{32}P-labeled cytosol; lane 4, 100 μl of ^{32}P-labeled L-cell cytosol mixed with 300 μl of [^{3}H]-dexamethasone mesylate-labeled cytosol before addition of the monoclonal antibody. The gel was visualized by autoradiography using EN^{3}HANCE. (From ref. *16*).

with [^{3}H]-dexamethasone mesylate, or ^{32}P-labeled cytosol, or a mixture of the two preparations were incubated with the monoclonal antibody and protein-A-Sepharose. Although it was difficult to achieve the optimal proportions of each labeled cytosol to react with antibody and put on the gel lane, there is a line of continuity between the [^{3}H]-dexamethasone mesylate-labeled band in lane 2, a ^{32}P-labeled band in lane 3, and the upper band in the doublet recovered from the mixture in lane 4. Again, by this method the dexamethasone mesylate-labeled protein is distinct from the 90–92K phosphoprotein.

As the 90–92K, non-steroid-binding phosphoprotein copurifies with the receptor by two independent methods, one of which is based on recognizing a steroid-binding site and the other on recognizing an antibody-binding site, we would propose that the 90–92K protein may be a component of the molybdate-stabilized, untransformed glucocorticoid-receptor complex in L-cell cytosol. As we find some of the 90–92K phosphoprotein under all of our conditions, we would suggest that it is present in cytosol in rather large amounts. Preincubation of cytosol with a receptor-saturating concentration of triamcinolone acetonide prior to affinity chromatography does not completely eliminate the ^{32}P-labeled 90–92K band (Fig. 7) and a small amount of 90–92K protein is bound to protein-A-Sepharose when non-immune mouse IgG or even no IgG is present (Fig. 8). A similar non-steroid-binding 90K phosphoprotein has been found in association with the 110K and 79K progesterone-binding phosphoproteins in molybdate-stabilized cytosol from chick oviduct (*6*, *7*, *26*, *28*, *37*).

IV. PHOSPHORYLATION OF THE GLUCOCORTICOID RECEPTOR IN A CELL-FREE SYSTEM

Recently, we (Sanchez and Pratt, work in progress) have been using a different ^{32}P-labeling approach to identify potential components of the molybdate-stabilized glucocorticoid receptor. L-cell cytosol is incubated on ice for 16 hr with polyclonal rabbit antiserum raised against the L-cell glucocorticoid receptor purified by affinity chromatography. The antibody is then adsorbed to protein-A-Sepharose, the protein-A-Sepharose pellet is washed, and the washed pellet is incubated at 25°C in buffer containing molybdate, dithiothreitol, and [γ-^{32}P]-ATP. After a second washing to remove unreacted ATP, the proteins are resolved by SDS-polyacrylamide gel electrophoresis. As

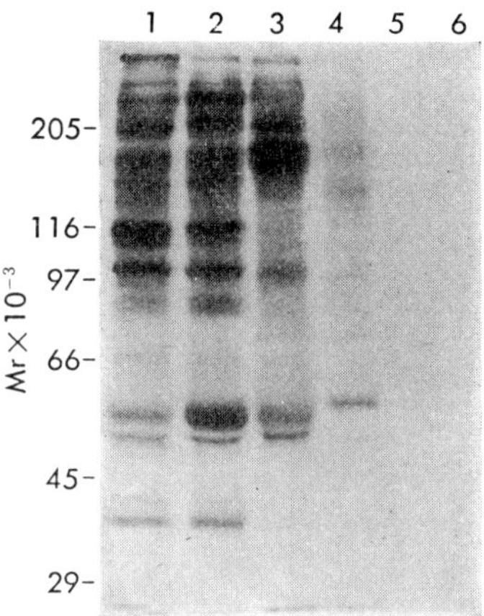

Fig. 10. Phosphorylation of the 98–100K protein and a 116K protein after molybdate-stabilized, L-cell cytosol proteins have been bound to protein-A-Sepharose in an immune specific manner using a polyclonal rabbit antiserum to the murine glucocorticoid receptor. L-cell cytosol (300 μl) diluted 1:1 with TEG buffer containing 20 mM sodium molybdate and 20 mM dithiothreitol was incubated on ice for 16 hr with 60 μl of immune or preimmune serum from a rabbit that had been immunized with L-cell glucocorticoid receptor purified by affinity chromatography on deoxycorticosterone-agarose. After incubation with antiserum, IgG was adsorbed to 150 μl of protein-A-Sepharose for 2 hr and washed in essentially the same manner as for the experiments of Figs. 8 and 9. The washed protein-A-Sepharose pellets were incubated for 30 min at 25°C with 116 μM [γ-^{32}P]-ATP, 4 mM $MgCl_2$ in TEG buffer containing 20 mM sodium molybdate and 20 mM dithiothreitol. After washing, the proteins were eluted with SDS buffer and resolved by electrophoresis on 7% SDS-polyacrylamide gels. Lanes 1 and 2 represent two different preparations of L-cell cytosol incubated with immune serum. Lane 3, cytosol incubated with preimmune serum. Lane 4, L-cell cytosol that was not incubated with either preimmune or immune serum prior to incubation with protein-A-Sepharose. Lanes 5 and 6 represent immune and preimmune serum, respectively. incubated with protein-A-Sepharose in the absence of L-cell cytosol. (From Sanchez and Pratt, work in progress).

shown in lanes 1 and 2 of Fig. 10, the material adsorbed to protein-A-Sepharose in the presence of the polyclonal antibody contains major ^{32}P-labeled bands at 116K, 98–100K, and 37K that are not phosphorylated in the protein-A-Sepharose pellet of cytosol preincubated with preimmune serum. It is clear from lanes 4, 5, and 6 of Fig. 10 that the kinase responsible for the phosphorylation is derived from the L-cell cytosol.

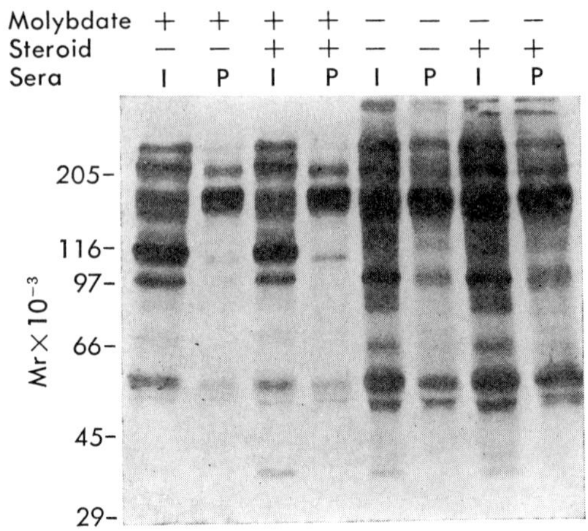

Fig. 11. The 116K protein is only identified in immunoabsorbed material from molybdate-stabilized cytosol, whereas phosphorylation of the 98–100K protein is independent of the presence of molybdate. Aliquots (300 μl) of L-cell cytosol containing molybdate (20 mM) and/or dexamethasone (50 nM) were incubated with immune or preimmune polyclonal rabbit antiserum, bound to protein-A-Sepharose, and incubated with [γ-^{32}P]-ATP and magnesium chloride as described in the legend to Fig. 10. The presence (+) or absence (−) of molybdate or steroid is indicated above each gel lane, as is the use of immune (I) or preimmune (P) antiserum. (From Sanchez and Pratt, work in progress).

In the experiment of Fig. 11 we asked whether the presence of either molybdate or steroid in the cytosol affected the profile of proteins that were phosphorylated. It is clear that the 98–100K protein is brought down in an immune-specific manner and is phosphorylated regardless of whether molybdate is present in the cytosol or not. The 116K protein is phosphorylated only in samples derived from molybdate-stabilized cytosol. The absence of the 116K band in samples obtained from cytosol that did not contain molybdate strongly suggests that the immune serum does not contain antibodies against the 116K protein. Thus, the 116K species could represent a component of the molybdate-stabilized, untransformed receptor that would not be seen except for its ability to be a substrate for whatever kinase enzyme(s) is in the material adsorbed to protein-A-Sepharose.

When we repeated the same experiment with the monoclonal antibody, we found that there was almost no phosphorylation of the 98–100K protein and no phosphorylation of the 116K band. This result could be due to the

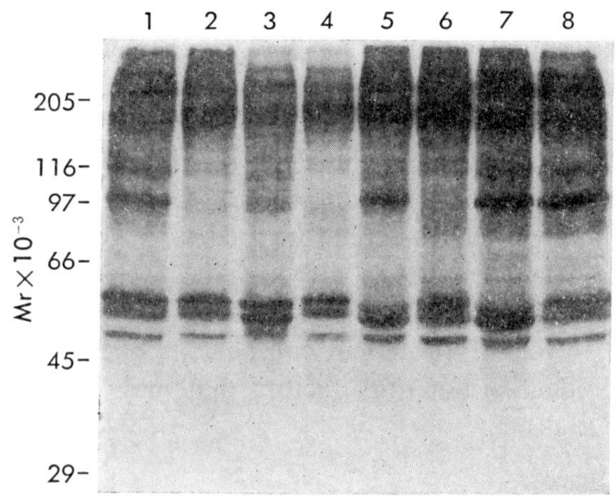

Fig. 12. Phosphorylation of the 98–100K protein after it has reacted with a monoclonal anti-receptor antibody and been adsorbed to protein-A-Sepharose. Aliquots (200 μl) of L-cell cytosol were incubated with 4 μl rabbit polyclonal or preimmune serum in the presence or absence of 20 μl monoclonal antibody or non-immune IgG and all samples were bound to protein-A-Sepharose. After washing, the protein-A-Sepharose-bound material was incubated with [γ-^{32}P]-ATP and magnesium chloride as described in the legend to Fig. 10. Lane 1, cytosol incubated with polyclonal antiserum alone; lane 2, cytosol incubated with preimmune serum alone; lane 3, cytosol incubated with monoclonal antibody alone; lane 4, cytosol incubated with non-immune IgG alone; lane 5, cytosol incubated with both monoclonal antibody and rabbit preimmune serum; lane 6, cytosol incubated with both non-immune mouse IgG and rabbit preimmune serum; lane 7, cytosol incubated with both monoclonal antibody and rabbit polyclonal antiserum; lane 8, cytosol incubated with both non-immune mouse IgG and rabbit polyclonal antiserum. (From Sanchez and Pratt, work in progress).

fact that the protein-A-Sepharose-bound material obtained from cytosol that had been incubated with the monoclonal antibody simply does not have the appropriate kinase activity. Therefore, we performed the mixed antibody experiment shown in Fig. 12. As one can see from lane 2, the protein-A-Sepharose pellet obtained from molybdate-stabilized cytosol incubated with preimmune rabbit serum has kinase activity, but no 116K or 98–100K proteins are phosphorylated. The protein-A-Sepharose pellet derived from molybdate-stabilized cytosol incubated with both the preimmune rabbit serum as a source of kinase activity and monoclonal antibody as a source of receptor and associated proteins (lane 5) contains a prominent phosphorylated band in the region of the receptor at 98–100K. There is no phosphorylation at 116K, however, suggesting that this protein either dissociates from the monoclonal antibody-receptor complex or that the 116K protein is present on the basis of an artifact that is unique to rabbit immune serum (as opposed to preimmune serum from the same rabbit and as opposed to mouse monoclonal serum that clearly recognizes the receptor). It is interesting that none of the various protein-A-Sepharose pellets have yielded much phosphorylation of the 90–92K protein, although it is clearly present by Coomassie Blue staining. These gels are presented to stimulate an interest in using this approach to identify potential components of molybdate-stabilized receptors and to identify potential factors that may be required for specific steroid binding to occur.

V. A MODEL DESCRIBING THE POTENTIAL ROLE OF PHOSPHATE AND SULFHYDRYL MOIETIES IN DETERMINING THE STEROID-BINDING CONFORMATION OF THE RECEPTOR

Several years ago, while studying the inactivation and activation of glucocorticoid-binding capacity in rat thymic lymphocyte cytosol, we found that molybdate stabilizes glucocorticoid receptors even when no binding activity is present (*29*). In the experiment shown in Fig. 13 rat thymocyte cytosol was incubated at 25°C in buffer alone, in buffer containing 2 mM dithiothreitol, or in buffer containing 10 mM molybdate. At various times, aliquots of each incubation were removed and the receptor activity was assayed by incubating each aliquot with [^{3}H]-triamcinolone acetonide at 0°C in the presence and absence of competing unlabeled steroid. As shown in panel A, addition of molybdate during the receptor inactivation occurring in the presence of dithiothreitol results in stabilization of whatever binding capacity is present at the time of molybdate addition. The addition of more dithiothreitol with the molybdate does not yield any more binding capacity than addition of only molybdate, showing that the receptor inactivation occurring in the presence of dithiothreitol is not merely a function of loss of dithiothreitol from the system. Part B of Fig. 13 shows that addition of dithiothreitol after varying times of receptor inactivation occurring in the absence of dithiothreitol and molybdate results in activation of the binding capacity to the level which exists in samples to which dithiothreitol was added at time zero. Part C of Fig. 13 shows that if thymocyte glucocorticoid receptors are in-

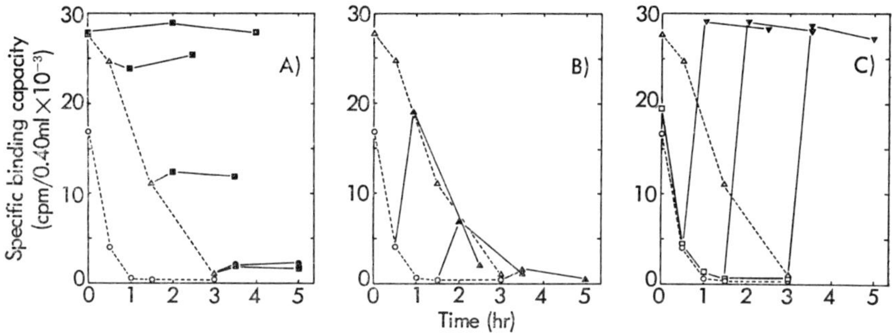

Fig. 13. Effects of molybdate and dithiothreitol on the stability of receptors in cytosol from rat thymic lymphocytes. Thymocyte cytosol was divided into several portions and incubated at 25°C under the following conditions. A: ○, control incubation in buffer without additions; △, incubation in the presence of 2 mM dithiothreitol; ■, dithiothreitol-containing samples to which 10 mM molybdate was added at the indicated times; ●, samples to which both molybdate and additional dithiothreitol were added after they had been incubated in the presence of dithiothreitol for 3 hr. In B, the control and dithiothreitol curves are repeated, and ▲ shows the binding capacity in samples that were exposed to dithiothreitol after incubation for 0.5, 1.5, or 3 hr in the presence of buffer alone. In C, □ represents the binding capacity in samples containing 10 mM molybdate and ▼ represents molybdate-containing samples to which dithiothreitol was added at 0.5, 1.5, or 3 hr. All values represent the average of duplicate assays expressed as counts per min/0.4 ml of incubation. (From ref. *29*).

activated in the presence of molybdate, which, as shown, does not stabilize the binding activity by itself, addition of dithiothreitol results in total reactivation of the binding capacity to the maximum time zero value no matter how little active receptor remains at the time of addition.

Two processes seem to be responsible for inactivation of the receptor in this experiment. The presence of dithiothreitol inhibits inactivation of the receptor by sulfhydryl oxidation, but even when adequate reducing agent is present, the receptor eventually becomes completely inactivated. Molybdate stabilizes the receptor even when no binding activity is present. Thus, molybdate appears to prevent a second inactivation process that occurs whether the sulfhydryl reducing agent is present or not. This observation that molybdate stabilizes receptors even when they are in a nonbinding state has been exploited in a second series of experiments that may eventually permit some insight into the mechanism of molybdate action. The experiments were

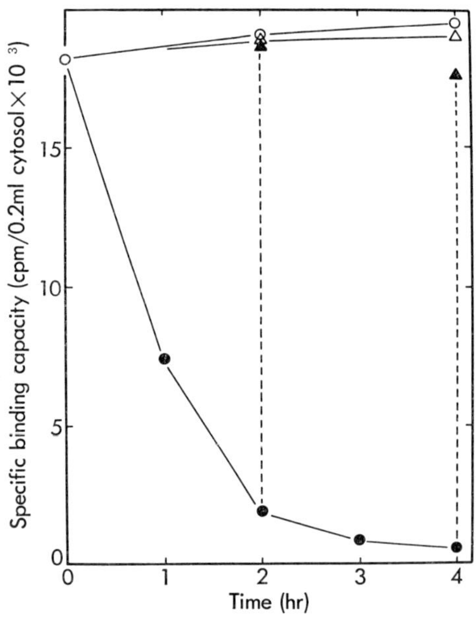

Fig. 14. Reactivation of phosphatase-treated receptors by dithiothreitol. Filtered rat liver cytosol containing molybdate was incubated alone (○), with purified calf intestine alkaline phosphatase at 20 μg/ml (●), or with alkaline phosphatase at 20 μg/ml plus 10 mM dithiothreitol (△). At the indicated times, 0.2 ml aliquots were removed and assayed for specific glucocorticoid-binding capacity. At 2 and 4 hr, 0.2 ml aliquots were removed from the incubation mixture containing alkaline phosphatase alone, dithiothreitol was added to a final concentration of 10 mM, and the specific binding capacity was assayed. The dithiothreitol-reactivated samples (▲) are connected to their corresponding phosphatase-inactivated samples by dashed lines. (From ref. *12*).

suggested by the finding that 10 mM molybdate does not inhibit calf intestine alkaline phosphatase activity assayed with *p*-nitrophenyl phosphate as substrate, and it does not prevent alkaline phosphatase-mediated inactivation of the glucocorticoid receptor (*19*). We reasoned that although molybdate might not be inhibiting inactivation of the glucocorticoid receptor by dephosphorylation, it might nevertheless stabilize the dephosphoreceptor such that it could be reactivated. Thus, we carried out the experiment shown in Fig. 14 (*12*).

Rat liver cytosol was first filtered and washed free of low Mr compounds

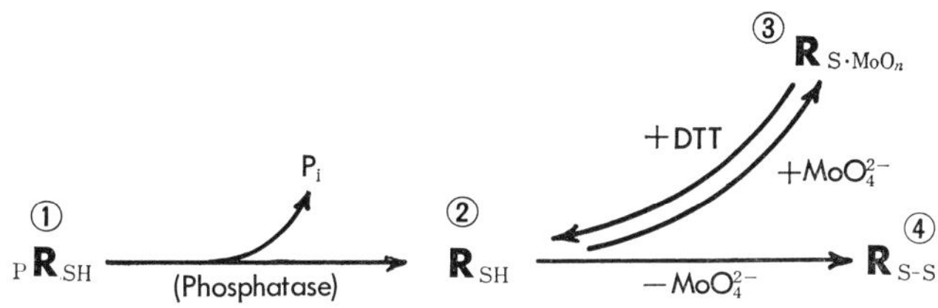

Fig. 15. Hypothetical model describing the different binding states of the receptor that exist in the experiment of Fig. 14. Receptor in states 1 and 2 binds steroid; state 3 is nonbinding, but reversible by reduction; and state 4 is nonbinding and not reversible by reduction. Molybdate is designated MoO_n when it is interacting with the receptor since the valence state is unknown under these conditions. The rationale for the model is described in the text.

that might inhibit phosphatase activity. Receptors were then inactivated in the presence of 10 mM molybdate by incubation with calf intestine alkaline phosphatase that had been purified to homogeneity. We found that the phosphatase-inactivated receptor can be restored to a steroid-binding state by the addition of dithiothreitol. Molybdate must be present during the inactivation for reactivation with dithiothreitol to occur. As the presence of dithiothreitol alone (*i.e.*, in the absence of molybdate) does not inhibit inactivation of binding capacity by the phosphatase, it is highly unlikely that the phosphatase is simply inactivating the endogenous receptor reducing system. The binding capacity can also be inactivated by highly purified rabbit muscle phosphoprotein phosphatase in the presence of molybdate, and subsequently reactivated in a similar manner (*12*).

Several potential binding states exist in this experiment and they are presented in the form of a model in Fig. 15. There is the untreated, active binding form, which is represented in the model by the phosphorylated and reduced receptor in state 1. Incubating cytosol with phosphatase in the presence of molybdate produces a nonbinding form of the receptor (state 3) that can be returned to a steroid-binding form (state 2) by addition of the reducing agent. As reactivation occurs at 0°C in the absence of ATP, phosphorylation is not involved and some kind of reduction must be occurring. If molybdate is not present, incubation with protein phosphatases results in a nonbinding state (state 4) that is not reactivated by reduction. As the binding capacity of cytosol is being inactivated by dephosphorylation and reactivated by reduction, we infer that dephosphorylation in some way promotes oxidation of sulfhydryl groups that are essential for steroid binding. It is reasonable to propose that a phosphate moiety(ies) on the receptor or a receptor-associated protein may be stabilizing vicinal thiol residues to maintain the active steroid-binding conformation. It is possible that, as defined for some other molybdate-protein interactions (*38*), the transition metal oxyanion complexes with sulfur groups on the receptor, preventing some change that produces a form of the receptor that cannot be reactivated by dithiothreitol. As the molybdate-stabilized, phosphatase-inactivated receptor can be reactivated, it would seem that the phosphate moiety(ies) itself is not required for steroid binding. Rather, it would seem that phosphorylation of the receptor may ensure a conformation such that sulfhydryl oxidation does not readily occur. Such a model is clearly speculative but it provides a conceptual framework for consideration of different binding states that appear to be determined by phosphate and sulfhydryl moieties on the receptor and for considering a potential mechanism for the stabilizing action of molybdate on steroid receptors.

SUMMARY

Glucocorticoid receptors are subject to two types of covalent modification —phosphorylation and reversible oxidation and reduction of sulfur moieties. Oxidation of the unbound receptor inactivates it with respect to its ability to bind steroid. Oxidized receptors can be returned to the steroid-binding state by addition of sulfhydryl reducing reagents or by incubation with boiled rat liver cytosol in the presence of molybdate. As receptor reduction is absolutely required for steroid binding, we have defined the endogenous heat-stable receptor-reducing system of rat liver cytosol. Two components of boiled cytosol are required to convert oxidized receptors to the steroid-binding state —NADPH and thioredoxin. Receptor reduction by boiled cytosol is inhibited in an immune-specific manner by antiserum to thioredoxin reductase. This leads us to propose that glucocorticoid receptors may be maintained in their active, steroid-binding conformation by an NADPH-dependent, thioredoxin-mediated reducing system.

The glucocorticoid receptor in L-cell cytosol has been demonstrated to be a 98–100K protein by site-specific affinity labeling with [^{3}H]-dexamethasone 21-mesylate. Cytosol prepared from L-cells that have been cultured for 18 hr in the presence of [^{32}P]-orthophosphate contains a 98–100K phosphoprotein and a 90–92K phosphoprotein that copurify by affinity chromatography on deoxycorticosterone-agarose. When ^{32}P-labeled L-cell cytosol is incubated with a monoclonal antibody against the glucocorticoid receptor and then adsorbed to protein-A-Sepharose, both the 90–92K non-steroid-binding phosphoprotein and the 98–100K steroid-binding phosphoprotein are bound to the protein-A-Sepharose pellet. As the 90–92K, non-steroid-binding phosphoprotein copurifies with the molybdate-stabilized receptor by two independent methods, we would propose that it may be a component of the untransformed glucocorticoid-receptor complex in L-cell cytosol.

As yet, there is no direct evidence that receptor phosphorylation is required for glucocorticoid binding. The results of some indirect experiments in which the glucocorticoid-binding capacity of rat liver cytosol is inactivated by purified protein phosphatases have suggested a model (shown in Fig. 15) in which phosphorylation of the receptor may ensure a conformation such that inactivation of steroid-binding capacity by sulfhydryl oxidation does not readily occur.

Acknowledgments
The authors' studies on receptor phosphorylation are supported by grant CA-28010 from the National Cancer Institute and studies on the endogenous receptor reducing system are supported by grant AM31573 from the National Institute of Arthritis, Diabetes, Digestive and Kidney Diseases.

REFERENCES

1) Bell, P.A. and Munck, A. (1973) Steroid-binding properties and stabilization of cytoplasmic glucocorticoid receptors from rat thymus cells. *Biochem. J.* **136**, 97–107.
2) Carter-Su, C. and Pratt, W.B. (1985) Receptor phosphorylation. In "The Receptors," Vol. 1, ed. Conn, P.M., Academic Press, New York, in press.
3) Chader, G.J. (1973) Some factors affecting the uptake, binding and retention of [^{3}H]-cortisol by the chick embryo retina as related to enzyme induction. *J. Neurochem.* **21**, 1525–1532.
4) Chang, L.M.S. and Bollum, F.J. (1981) Immunological reagents for comparisons of DNA polymerase-α and DNA polymerase-β. *J. Biol. Chem.* **256**, 494–498.
5) Dahmer, M.K., Housley, P.R., and Pratt, W.B. (1984) Effects of molybdate

and endogenous inhibitors on steroid receptor inactivation, transformation and translocation. *Annu. Rev. Physiol.* **46**, 67–81.

6) Dougherty, J.J., Puri, R.K., and Toft, D.O. (1984) Polypeptide components of two 8S forms of chicken oviduct progesterone receptor. *J. Biol. Chem.* in press.
7) Gasc, J.M., Renoir, J.M., Radanyi, C., Joab, I., Tuohimaa, P., and Baulieu, E.E. (1984) Progesterone receptor in chick oviduct: An immunohistochemical study with antibodies to distinct receptor components. *J. Cell Biol.* **99**, 1193–1201.
8) Granberg, J.P. and Ballard, P.L. (1977) The role of sulfhydryl groups in the binding of glucocorticoids by cytoplasmic receptors of lung and other mammalian tisues. *Endocrinology* **100**, 1160–1168.
9) Grippo, J.F., Tienrungroj, W., Dahmer, M.K., Housley, P.R., and Pratt, W.B. (1983) Evidence that the endogenous heat-stable glucocorticoid receptor activating factor is thioredoxin. *J. Biol. Chem.*, **258**, 13658–13664.
10) Grippo, J.F., Holmgren, A., and Pratt, W.B. (1985) Proof that the endogenous heat-stable glucocorticoid receptor activating factor is thioredoxin. *J. Biol. Chem.* **260**, in press.
11) Holmgren, A. (1979) Thioredoxin catalyzes the reduction of insulin disulfides by dithiothreitol and dihydrolipoamide. *J. Biol. Chem.* **254**, 9627–9632.
12) Housley, P.R., Dahmer, M.K., and Pratt, W.B. (1982) Inactivation of glucocorticoid-binding capacity by protein phosphatases in the presence of molybdate and complete reactivation by dithiothreitol. *J. Biol. Chem.* **257**, 8615–8618.
13) Housley, P.R. and Pratt, W.B. (1983) Direct demonstration of glucocorticoid receptor phosphorylation by intact L-cells. *J. Biol. Chem.* **258**, 4630–4635.
14) Housley, P.R., Grippo, J.F., Dahmer, M.K., and Pratt, W.B. (1984) Inactivation, activation and stabilization of glucocorticoid receptors. In "Biochemical Actions of Hormones," ed. Litwack, G., pp. 347–376, Academic Press, New York.
15) Housley, P.R., Grippo, J.F., and Pratt, W.B. (1985) Steroid receptor phosphorylation. In "Hormone Responsive Tumors," ed. Hollander, C.P., Academic Press, New York, in press.
16) Housley, P.R., Sanchez, E.R., Westphal, H.M., Beato, M., and Pratt, W.B. (1985) The molybdate-stabilized L-cell glucocorticoid receptor isolated by affinity chromatography or with a monoclonal antibody is associated with a 90–92K non-steroid-binding phosphoprotein. *J. Biol. Chem.*, in press.
17) Ishii, D.N., Pratt, W.B., and Aronow, L. (1972) Steady-state level of the specific binding component of mouse fibroblasts. *Biochemistry* **11**, 3896–3904.
18) Lan, N.C., Wang, T.S.F., Johnson, L.K., and Korn, D. (1984) Monoclonal

antibodies against human DNA polymerase-α do not cross-react with glucocorticoid receptors. *J. Steroid Biochem.* **20**, 251–254.

19) Leach, K.L., Dahmer, M.D., and Pratt, W.B. (1983) Glucocorticoid receptor stabilization: Relative effects of molybdate ion on inactivation by alkaline phosphatase and phospholipase A_2. *J. Steroid Biochem.* **18**, 105–107.
20) Luthman, M. and Holmgren, A. (1982) Rat liver thioredoxin and thioredoxin reductase: Purification and characterization. *Biochemistry* **21**, 6628–6633.
21) Mark, D. and Richardson, C.C. (1976) *Escherichia coli* thioredoxin: A subunit of bacteriophage T7 DNA polymerase. *Proc. Natl. Acad. Sci. U.S.A.* **73**, 780–784.
22) Munck, A. and Brink-Johnsen, T. (1968) Specific and nonspecific physicochemical interactions of glucocorticoids and related steroids with rat thymus cells *in vitro*. *J. Biol. Chem.* **243**, 5556–5565.
23) Munck, A., Wira, C., Young, D.A., Mosher, K.M., Hallahan, C., and Bell, P.A. (1972) Glucocorticoid-receptor complexes and the earliest steps in the action of glucocorticoids on thymus cells. *J. Steroid. Biochem.* **3**, 567–578.
24) Nordström, B., Randahl, H., Slaby, I., and Holmgren, A. (1981) Characterization of bacteriophage T7 DNA polymerase purified to homogeneity by antithioredoxin immunoadsorbent chromatography. *J. Biol. Chem.* **256**, 3112–3117.
25) Pratt, W.B., Kaine, J.L., and Pratt, D.V. (1975) The kinetics of glucocorticoid binding to the soluble specific binding protein of mouse fibroblasts. *J. Biol. Chem.* **250**, 4584–4591.
26) Radanyi, C., Joab, I., Renoir, J.M., Richard-Foy, H., and Baulieu, E.E. (1983) Monoclonal antibody to chicken oviduct progesterone receptor. *Proc. Natl. Acad. Sci. U.S.A.* **80**, 2854–2858.
27) Rees, A.M. and Bell, P.A. (1975) The involvement of receptor sulfhydryl groups in the binding of steroids to the cytoplasmic glucocorticoid receptor from rat thymus. *Biochim. Biophys. Acta* **411**, 121–132.
28) Renoir, J.M., Mester, J., Buchou, T., Catelli, M.G., Touhimaa, P., Binart, N., Joab, I., Radanyi, C., and Baulieu, E.E. (1984) Purification by affinity chromatography and immunological characterization of a 110K Da component of the chick oviduct progesterone receptor. *Biochem. J.* **217**, 685–692.
29) Sando, J.J., Hammond, N.D., Stratford, C.A., and Pratt, W.B. (1979) Activation of thymocyte glucocorticoid receptors to the steroid-binding form. The roles of reducing agents, ATP, and heat-stable factors. *J. Biol. Chem.* **254**, 4779–4789.
30) Schmidt, T.J., Bollum, F.J., and Litwack, G. (1982) Correlations between the activities of DNA polymerase α and the glucocorticoid receptor. *Proc. Natl. Acad. Sci. U.S.A.* **79**, 4555–4559.

31) Simons, S.S., Jr., Pons, M., and Johnson, D.F. (1980) α-Keto mesylate: A reactive, thiol-specific functional group. *J. Org. Chem.* **45**, 3084–3088.
32) Simons, S.S., Jr., and Thompson, E.B. (1981) Dexamethasone 21-mesylate: An affinity label of glucocorticoid receptors from rat hepatoma tissue culture cells. *Proc. Natl. Acad. Sci. U.S.A.* **78**, 3541–3545.
33) Sloman, J.C. and Bell, P.A. (1976) The dependence of specific nuclear binding of glucocorticoid by rat thymus cells on cellular ATP. *Biochim. Biophys. Acta* **428**, 403–413.
34) Tanaka, S., Hu, S.Z., Wang, T.S.F., and Korn, D. (1982) Preparation and preliminary characterization of monoclonal antibodies against human DNA polymerase α. *J. Biol. Chem.* **257**, 8386–8390.
35) Thelander, L. (1974) Reaction mechanism of ribonucleoside diphosphate reductase from *Escherichia coli* oxidation-reduction-active disulfides in the B1 subunit. *J. Biol. Chem.* **249**, 4858–4862.
36) Thelander, L. and Reichard, P. (1979) Reduction of ribonucleotides. *Annu. Rev. Biochem.* **48**, 133–158.
37) Touhimaa, P., Renoir, J.M., Radanyi, C., Mester, J., Joab, I., Buchou, T., and Baulieu, E.E. (1984) Antibodies against highly purified B-subunit of the chick oviduct progesterone receptor. *Biochem. Biophys. Res. Commun.* **119**, 433–439.
38) Weathers, B.J., Grate, J.H., and Schrauzer, G.N. (1979) The chemical evolution of a nitrogenase model. 17. Stimulation of stearic and of inhibitory effects at the enzyme active site with acetylene and nitriles as the substrates and "molybdoinsulin" catalysts. *J. Am. Chem. Soc.* **101**, 917–924.
39) Westphal, H.M., Moldenhauer, G., and Beato, M. (1982) Monoclonal antibodies to the rat liver glucocorticoid receptor. *EMBO J.* **1**, 1467–1471.
40) Wheeler, R.H., Leach, K.L., LaForest, A.C., O'Toole, T.E., Wagner, R., and Pratt, W.B. (1981) Glucocorticoid receptor activation and inactivation in cultured human lymphocytes. *J. Biol. Chem* **256**, 434–441.

2

Glucocorticoid Hormone: Mechanisms of Action, ed. Sakamoto, Y. and Isohashi, F., pp. 35–66,
Japan Sci. Soc. Press, Tokyo/Springer-Verlag, Berlin (1986)

PURIFICATION, ACTIVATION, AND PHOSPHORYLATION OF THE GLUCOCORTICOID RECEPTOR

THOMAS J. SCHMIDT*[1] and GERALD LITWACK*[2]

Department of Physiology and Biophysics, College of Medicine, The University of Iowa, Iowa City, Iowa 52242[1] and Fels Research Institute and Department of Biochemistry, Temple University School of Medicine, Philadelphia, Pennsylvania 19140*[2], U.S.A.*

Glucocorticoids influence the growth, differentiation, and function of a wide variety of tissues and cell types. These responses vary dramatically and include catabolic effects, such as lysis of lymphoid cells (*18*, *70*, *75*), as well as anabolic effects, including induction of a limited number of enzymes in hepatoma cells (*9*, *90*, *91*). Extensive experimental data support the concept that, as with other steroid hormones, most of these effects of glucocorticoids are mediated through intracellular receptor proteins which bind this steroid molecule with high affinity and specificity (*67*). Published studies have clearly suggested that this binding requires phosphorylation as well as reduction of the receptor protein itself or of some other cytosolic component required for the steroid-binding reaction (*79*, *80*). If these criteria are met and the glucocorticoid molecule binds to its cytoplasmic receptor, the complex must then undergo a two-step process in order to bind to nuclei and ultimately affect gene expression. The first temperature-dependent step, termed "activation" or "transformation," is thought to involve a conformational change in the steroid-receptor complex resulting in the exposure of positively charged amino acid residues (consisting minimally of arginine, histidine, and lysine) on the surface of the protein (*19*, *61*) and hence an increased affinity for nuclei and polyanions such as purified DNA and DNA-cellulose (*11*, *48*, *52*, *61*). This conformational change is also reflected in an altered elution profile of the activated glucocorticoid-receptor complexes from anion exchange resins such as DEAE-Sephadex (*72*) and DEAE-cellulose (*78*). The second temperature-independent step which follows activation is termed "translocation" and involves the movement of the previously activated complexes into the nucleus where they bind to acceptor sites within the chromatin (*13*) and, more precisely, to DNA sequences which flank specific genes whose expression is regulated by glucocorticoids (*29*, *74*, *81*). The net results of this nuclear binding are the initiation of transcription of these glucocorticoid-responsive genes and the ultimate translation of mRNAs into specific proteins whose activities contribute to the phenotypic response.

Maximal *in vitro* activation of glucocorticoid-receptor complexes in crude cytosolic extracts can be achieved by a number of manipulations including elevated temperature, increased ionic strength, elevated pH, or gel filtration (*6*, *27*, *61*). However, the precise biochemical mechanism(s) underlying this crucial step remains to be elucidated. Numerous enzymatic and non-enzymatic theories including dephosphorylation of the receptor itself or an accessory protein (*8*, *50*), glucocorticoid-receptor phosphorylation (*47*), limited proteolysis (*63*), dissociation of a low molecular weight inhibitor or "modulator" (*5*, *27*, *85*), and simple subunit dissociation (*97*) have all been proposed as possible mechanisms involved in activation. However, rigorous testing of these theories, which have been evaluated in several published reviews (*36*, *60*, *83*), requires purification of unactivated glucocorticoid-receptor complexes and re-evaluation of the *in vitro* regulation of activation and the potential role of cytoplasmic factors in mediating this conformational change. Since activation of glucocorticoid-receptor complexes occurs *in vivo* under physiological conditions and is rate-limiting for nuclear binding (*57*, *64*, *66*), insight with respect to the biochemical mechanism underlying this obligatory step would contribute to our understanding of how hormonal signals are transmitted within the target cell.

Data concerning both the structure of the receptor as well as the mechanism of activation have been generated primarily with crude cytoplasmic extracts and hence are subject to multiple interpretations. Several laboratories have therefore attempted to develop a suitable purification scheme for this protein. Early attempts to purify the glucocorticoid-receptor complex in its unactivated form were relatively unsuccessful for several reasons. First, the steroid-binding site of this protein, which is present at very low concentrations, is relatively unstable and secondly, most of the techniques used to purify proteins, including ammonium sulfate precipitation and gel filtration, quickly activate the bound glucocorticoid-receptor complexes. However, one of the first successful attempts at purifying unactivated complexes utilized affinity chromatography. Failla *et al.* (*24*) reported the partial purification (approximately 2000-fold) of unactivated glucocorticoid-receptor complexes from hepatoma tissue culture cells using a two-step procedure involving biospecific adsorption of the receptors to deoxycorticosterone derivatized agarose, elution with [^{3}H]-triamcinolone acetonide (TA) and subsequent gel filtration on Bio-Gel A-0.5 m. These partially purified complexes were judged to be unactivated because subsequent warming (20°C for 30 min) enhanced their nuclear binding capacity by approximately 2.6-fold. In these experiments the pH was maintained at 6.5, which probably minimized spontaneous activation which could have occurred during the purification procedure. Although additional reports describing purification of unactivated glucocorticoid receptors appeared (*28*, *55*), neither the physicochemical characteristics of the receptors or the mechanism of activation were thoroughly investigated. After Leach *et al.* (*50*) demonstrated that sodium molybdate blocks both the inactivation (loss of steroid-binding capacity) of unbound receptors as well as the activation of crude glucocorticoid-receptor complexes, the technical problem concerning spontaneous activation during purification was subsequently overcome by including this inhibitor in all purification buffers. Most recently a three-step purification scheme which includes affinity chromatography, gel filtration, and anion exchange chromatography (*33*) has been employed successfully in our laboratories to yield highly purified, molybdate-stabilized, unactivated glucocorticoid-receptor complexes. This purification scheme also involves the initial adsorption of unbound cytoplasmic receptors to a deoxycorticosterone-derivatized agarose developed by Grandics (*30*, *31*). As will be discussed later, this affinity resin offers specific advantages over similar resins utilized previously.

Experimental data generated in several laboratories suggest that phosphorylation/dephosphorylation reactions may regulate receptor functions including both glucocorticoid-binding activity as well as the subsequent activation step (*83*). However, since unpurified glucocorticoid receptors were used in these studies it was not possible to distinguish between phosphorylation/dephosphorylation of the receptor protein itself and similar modifications of other cellular components which influence receptor functions. Numerous published studies have established the fact that the glucocorticoid-binding ability of cells can be correlated with ATP levels (*14*, *65*, *102*) and that the specific cytosolic binding that decreases during *in vitro* incubations can be partially restored by ATP (*80*). Although these data are consistent with regenerated steroid binding resulting as a consequence of phosphorylation. Barnett *et al.* (*7*) have demonstrated that ADP generated from ATP may interact with a receptor nucleotide-binding site and stabilize the glucocorticoid-binding site by an allosteric mechanism rather than by phosphorylation. However, the hypothesis that dephosphorylation might regulate glucocorticoid binding is supported by the observations that incubation of crude cytosol with exogenous alkaline phosphatase greatly diminishes its capacity to bind glucocorticoids (*69*) and phosphatase inhibitors, such as molybdate and fluoride, stabilize the binding capacity of cytosolic glucocorticoid receptors (*50*, *80*). The greater effectiveness of purified rabbit muscle protein phosphatase (preferential activity toward phosphoserine residues) when compared to alkaline phosphatase (preferential activity toward phosphotyrosine residues) in mediating inactivation of unoccupied receptors (*41*) is consistent with the observation that fibroblast glucocorticoid receptors are phosphorylated *in vivo* on serine residues (*43*).

Likewise, the data which suggest that dephosphorylation may be involved in glucocorticoid-receptor activation is indirect. Exogenous alkaline phosphatase, which secondarily dephosphorylates phosphoproteins, stimulates the rate of activation of rat hepatic glucocorticoid-receptor complexes (*8*) and the phosphatase inhibitor, sodium molybdate, inhibits activation (*8*, *50*). However, it should be noted that molybdate may not block inactivation of unbound receptors and activation of bound complexes by simply inhibiting a phosphoprotein phosphatase which affects both processes. Harmon *et al.* (*38*) have reported that although sodium molybdate blocks inactivation of unbound receptors in an "activation labile," glucocorticoid resistant lymphoid mutant, it does not apparently inhibit activation of glucocorticoid-receptor complexes in this same mutant.

Finally, several laboratories in addition to our own have published data which demonstrate that the glucocorticoid receptor, like the progesterone (*20*,

(*26*, *101*) and estrogen receptors (*58*, *59*), is also a phosphoprotein. As previously mentioned, Housley and Pratt (*43*) have reported that receptors in glucocorticoid-sensitive mouse fibroblasts (L929) are phosphorylated *in vivo* on serine residues. As will be discussed later in this chapter, our laboratory has also demonstrated that rat hepatic glucocorticoid receptors are labeled *in vivo* with [^{32}P]-orthophosphate (*32*). Kurl and Jacob (*49*) have also published preliminary data which suggest that the purified rat hepatic glucocorticoid receptor contains protein kinase activity which is capable of phosphorylating the receptor. The purity of the receptor utilized for these experiments was based solely on staining with Coomassie Blue and hence minor contamination by a kinase which might have co-purified with the receptor could not be totally excluded. Neither the potential dependency of the kinase activity on ligand binding or on cyclic nucleotides were investigated. Before these data were published our laboratory had already initiated a thorough biochemical characterization of the kinase activity associated with the highly purified rat hepatic glucocorticoid receptor and those data will also be reviewed in this chapter.

I. RECEPTOR PURIFICATION

The three-step scheme (affinity chromatography, gel filtration, DEAE-cellulose chromatography) which our laboratory has developed for purification of unactivated rat hepatic glucocorticoid receptors has been described in detail elsewhere (*33*) and therefore only will be outlined briefly here. The livers of adrenalectomized male Wistar rats (150–200 g) are first perfused *in situ via* the portal vein with cold 0.9% NaCl and then with homogenization buffer A (50 mM potassium phosphate, 10 mM sodium molybdate, and 10 mM thioglycerol, pH 7.0 at 4°C). The excised and minced livers are then weighed and homogenized with a motor driven Teflon-glass Potter-Elvehjem apparatus in 1 volume of buffer A. After the crude homogenate is centrifuged at 4000 × *g* for 10 min and the upper lipid layer is carefully aspirated, the supernatant fraction is centrifuged at 105,000 × *g* for 1 hr to obtain the cytosol. All of these procedures are performed at 0–4°C and the final cytosol is stored at −70°C until further use. To initiate the purification procedure, approximately 10 ml of thawed, molybdate-containing cytosol is mixed with 5 ml of a deoxycorticosterone-derivatized agarose synthesized by Grandics *et al.* (*30*, *31*) and depicted in Fig. 1.

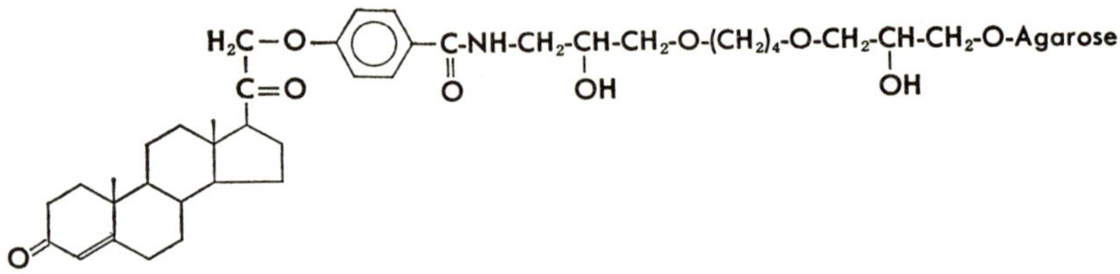

Fig. 1. Structure of affinity resin.

Because of the steps employed in its preparation, this resin offers several advantages for the purification of cytoplasmic glucocorticoid receptors (*30*). Although the cyanogen bromide technique (*4*, *56*) has frequently been used to attach various ligands to polysaccharides, this procedure often introduces anion exchange groups on the support (*103*) and hence enhances nonspecific protein adsorption. In addition, the immobilized ligand often leaks from the matrix due to the lability of the isourea bond formed during the introduction of the spacer (*73*). In contrast, adsorbents such as the present affinity resin, which have been prepared by the epoxide technique (*89*) usually exhibit much higher stability and lower nonspecific adsorption (*68*). In addition to these obvious advantages, other modifications were introduced during synthesis of the deoxycorticosterone-derivatized agarose (*30*). The reactive groups which were introduced into the polysaccharide support by the initial epoxidation step were converted into non-reactive amino groups and the epoxidation step itself was modified to prevent formation of extended crosslinks on the carrier (*89*).

After crude cytosol is stirred with this affinity resin at 0–4°C for a minimum of 2 hr, the supernatant fraction (referred to as the affinity gel supernatant) is collected by centrifugation and assayed for specific binding of [6,7-^{3}H]-TA (32–37 Ci/mmol; New England Nuclear) by the hydroxylapatite technique (*23*, *24*). The resin is then washed batchwise eight times with 10 ml of buffer A to remove nonspecifically adsorbed proteins. The receptors are then eluted from the resin by adding 5–10 ml of buffer B (10 mM potassium phosphate, 10 mM sodium molybdate, 10 mM thioglycerol, and 10% glycerol, pH 7.0 at 4°C) containing 2 μM [^{3}H]-TA and incubating the mixture at 0–4°C for 16 hr. The eluted [^{3}H]-TA-receptor complexes are then collected by centrifugation or by filtering the slurry on a Teflon filter. The resin is then washed with an additional 4 ml of buffer B and the combined affinity gel eluate is assayed for bound [^{3}H]-TA. The used affinity resin is then regenerated as previously described (*31*) and is transferred into buffer A and stored at 0–4°C. The affinity gel eluate is then filtered on a Bio-Gel A-1.5 m column (220 ml bed volume; inner diameter 2.5 cm) equilibrated with buffer A at 0–4°C. Eight-ml fractions are collected and aliquots are counted for radioactivity. The [^{3}H]-TA-receptor fractions elute as a single peak of bound radioactivity as seen in Fig. 2. The receptor peak fractions are then com-

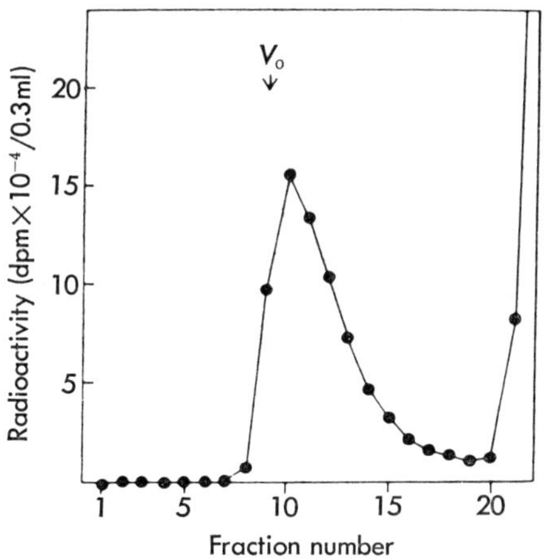

Fig. 2. Bio-Gel A-1.5 m column chromatography of affinity gel eluted [^{3}H]-TA-receptor complexes. Reproduced with permission from *J. Biol. Chem* (*33*).

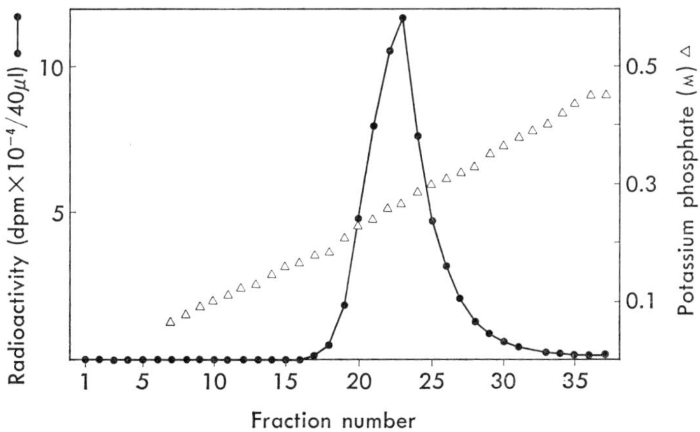

Fig. 3. DEAE-cellulose chromatography of purified [^{3}H]-TA-receptor complexes. Reproduced with permission from *J. Biol. Chem.* (*33*).

bined and applied *via* a peristaltic pump to a DEAE-cellulose column (3-ml bed volume) and the unactivated [^{3}H]-TA-receptor complexes are then eluted with a linear salt gradient (50–500 mM potassium phosphate) containing 10 mM sodium molybdate and 10 mM thioglycerol, pH 7.0. Thirty 1-ml fractions are collected and 50 μl aliquots are assayed for determination of radioactivity. As seen in Fig. 3, the unactivated [^{3}H]-TA-receptor complexes are eluted as a single peak at approximately 280 mM potassium phosphate. The unactivated [^{3}H]-TA-receptor peak fractions are then pooled and either utilized immediately for experimentation or frozen under liquid nitrogen for use at a later date. Prior to freezing, however, the pooled peak is filtered through a Sephadex G-25 column equilibrated with buffer C (10 mM 2(N-morpholino) ethanesulfonic acid (MES), 0.5 mM ethylenediaminetetraacetate (EDTA), 0.5 mM dithiothreitol, and 5 mg/ml of bovine serum albumin, pH 7.0 at 4°C) containing 10 mM MoO_4 to remove the potassium phosphate. The eluate is adjusted to 10 % glycerol and 60 nM [^{3}H]-TA is added to facilitate rebinding during the thawing process. After thawing, the purified, unactivated [^{3}H]-TA-receptor complexes are again filtered through a Sephadex G-25 column to facilitate removal of sodium molybdate, glycerol, and free [^{3}H]-TA prior to activation studies.

II. CHARACTERIZATION OF PURIFIED, UNACTIVATED RECEPTOR COMPLEXES

The unactivated, molybdate-stabilized rat hepatic glucocorticoid receptor has been purified approximately 4000-fold, as calculated by specific radioactivity, by the three-step procedure described (*33*). The average percent yield and the fold purification after the affinity gel, gel filtration, and DEAE-cellulose chromatography steps are listed in Table I. The purity of the receptor peak

TABLE I
Three-step Purification of Unactivated Rat Hepatic Glucocorticoid Receptors

Fraction	Yield (%)	Fold purification[a]
Crude cytosol	100	1
Affinity gel eluate	60	144
Bio-Gel filtrate	50	566
DEAE-cellulose eluate	34	4,295

[a] Average data for three separate experiments (*33*).

fractions eluted from DEAE-cellulose (Fig. 3) was analyzed by sodium-dodecylsulfate (SDS)-polyacrylamide gel electrophoresis (*71*) and the gels were subsequently subjected to the sensitive double staining procedure with Coomassie Blue R-250 followed by silver (*46*). The protein pattern revealed a major Mr=90,000 band in each fraction and the occurrence of this band closely paralleled the distribution of bound radioactivity seen in Fig. 3. This observation is consistent with the fact that a number of photoaffinity and electrophilic affinity labeling as well as immunological studies have identified a Mr=90,000 protein as the glucocorticoid receptor in a variety of tissues (*17*, *22*, *37*, *43*, *87*). The receptor peak fraction eluted from DEAE-cellulose was similarly electrophoresed and the gel was subjected to the double staining procedure, destained with Kodak Rapid Fix (*104*) and then restained by the double staining procedure. A scan of this gel in a Zeineh soft laser scanning densitometer revealed the presence of two other protein bands corresponding to apparent Mr=41,000 and 40,000. A Mr=24,000 band, which appeared to be present at very low concentrations in the receptor peak fraction, was also eluted from DEAE-cellulose at a salt concentration higher than that required for the elution of purified, unactivated [^{3}H]-TA-receptor complexes. The relationship of these three components (Mr=41,000, 40,000, and 24,000) to the Mr=90,000 receptor was investigated by presaturating liver cytosol with nonradioactive TA prior to proceeding with the purification procedure. When the appropriate fractions eluted from the final DEAE-cellulose column were analyzed, the Mr=90,000 component was very faint and the other three components were totally undetectable. These data suggested that these lower molecular weight components may either be constituents of the purified, molybdate-stabilized, unactivated glucocorticoid-receptor complex or may represent proteolytic fragments of the Mr=90,000 protein. However, as will be discussed later in this chapter, the lack of incorporation of ^{32}P into either the Mr=41,000 or 40,000 component suggests that they are not proteolytic fragments of the [^{32}P]-labeled Mr=90,000 receptor (*32*).

The highly purified unactivated [^{3}H]-TA-receptor complexes eluted from DEAE-cellulose sedimented at 9–10S in low salt (40 mM KCl) linear (10–35%) glycerol gradients containing 10 mM molybdate, and this sedimentation coefficient is similar to that reported for crude hepatic glucocorticoid receptors (*86*). However, when the salt concentration was increased to 1 M KCl the sedimentation coefficient of the purified [^{3}H]-TA-receptor complexes was only slightly reduced to 8–9S. Thus, unlike crude cytosolic receptors (*86*), the purified receptor complexes appear to be more resistant to a salt-induced conversion to an 8S form. Elevation of the salt concentration of the purified preparations also did not result in the appearance of 4–5S forms as previously detected with other molybdate-stabilized, crude cytosolic glucocorticoid receptors (*97*, *98*). These results suggest that the salt-induced conversion of the large (9–10S) receptor into smaller forms may not simply involve dissociation of subunits. As will be discussed shortly, a heat-stable cytosolic protein appears to facilitate maximal DNA-cellulose binding of thermally activated, purified [^{3}H]-TA-receptor complexes. It will be interesting to test whether this same factor might also play a role in generating the 4–5S form of the purified receptor.

The Stokes radius of the purified, unactivated [^{3}H]-TA-receptor complexes was estimated by filtration through a standardized column of Bio-Gel A-1.5 m agarose. The initial Stokes radius which was obtained was approximately 6.0 nm, but subsequent analyses suggested that dissociation of receptor complexes had occurred even in the presence of molybdate. This conclusion was consistent with the fact that multimeric proteins of extremely low concentration can dissociate upon gel filtration in a concentration-dependent manner (*1*, *2*) and the protein concentration in the purified receptor preparation was in the microgram/ml range. When a similar gel filtration column equilibrated with 50 μg/ml of albumin was utilized, a Stokes radius of 7.3 nm was calculated and this value is again similar to that of crude hepatic receptors (*86*). From the sedimentation coefficient and the Stokes radius an apparent Mr=303,000 was calculated for the purified receptor. A calculated frictional ratio of 1.64 and an axial ratio for a prolate ellipsoid of approximately 12 suggested that the purified receptor is highly asymmetric. Taken collectively, these data suggest that the hydrodynamic properties and molecular parameters of the highly purified complexes resemble those of molybdate-stabilized, crude hepatic glucocorticoid-receptor complexes (*86*).

III. ACTIVATION OF PURIFIED [^{3}H]-TA-RECEPTOR COMPLEXES

It was clear that the [^{3}H]-TA-receptor complexes which were purified by this three-step procedure were in the unactivated form because they could subsequently be heat activated and demonstrated enhanced binding to DNA-cellulose (*33*). In order to perform these activation experiments, however, the purified complexes were first filtered through Sephadex G-25 to remove

TABLE II
In Vitro Activation of Purified [^{3}H]-TA-Receptor Complexes Eluted from DEAE-cellulose

Treatment	Number of experiments	[^{3}H]-TA-receptor dpm added[a]	DNA-cellulose bound dpm	% of total bound to DNA-cellulose
Gel filtration in the	1	111,600	1,000	0.9
presence of 10 mM	2	29,700	270	0.9
Na_2MoO_4				
Gel filtration	1	86,900	6,200	7.2
	2	28,600	1,320	4.6
Gel filtration + heat	1	84,000	15,200	18.1
(25°C, 30 min)	2	24,100	4,470	18.5
Gel filtration + 10 mM	1	86,500	7,100	8.2
Na_2MoO_4 + heat	2	25,000	2,300	9.2
(25°C, 30 min)				
Gel filtration + cytosol +	1	85,400	39,000	45.6
heat (25°C, 30 min)	2	24,900	10,700	42.8
Gel filtration + cytosol +	1	69,500	9,800	14.1
10 mM Na_2MoO_4 + heat	2	25,500	2,960	11.6
(25°C, 30 min)				

[a] After each treatment bound [^{3}H]-TA was determined by the hydroxylapatite technique (*23*, *34*). Reproduced with permission from *J. Biol. Chem.* (*33*).

molybdate which was present throughout the purification procedure. As seen in Table II, this procedure alone resulted in increased binding of the purified [^{3}H]-TA-receptor complexes to DNA-cellulose, which agrees with the known activating effect of gel filtration (*6*, *27*). What was clear from these preliminary experiments was that although heating (25°C for 30 min) enhanced the binding of purified [^{3}H]-TA-receptor complexes to DNA-cellulose, this binding was significantly less than that previously observed for receptors in crude cytosols (*22*, *27*). In our laboratory generally 35–45% of the [^{3}H]-TA-receptor complexes in crude cytosols bind to DNA-cellulose after heating at this temperature (*8*, *84*). This observation prompted us to attempt thermal activation of purified complexes in a reconstituted system in which crude hepatic cytosol presaturated with nonradioactive TA was mixed with the purified, unactivated [^{3}H]-TA-receptor complexes. As seen in Table II, readdition of crude hepatic cytosol enhanced the DNA-cellulose binding of the purified [^{3}H]-TA-receptor complexes to a level similar to that detected with complexes in crude cytosols. These data clearly suggested that a cytoplasmic factor(s) may stimulate activation of [^{3}H]-TA-receptor complexes purified from rat liver. This same factor(s) appears to be present in rat kidney cortex cytosol and enhances the DNA-cellulose binding of [^{3}H]-TA-receptor complexes purified from that tissue (*99*). Since the publication of these preliminary activation data (*33*) we have attempted to further characterize this hepatic cytosolic "stimulator" and determine the mechanism(s) by which it apparently affects the conformation of the purified receptor. These more recent experiments (T. Schmidt, A. Miller-Diener, M. Webb, and G. Litwack, manuscript in preparation) will now be summarized and the conclusions drawn from these data will be briefly discussed.

The cytoplasmic "activating factor" or "stimulator" which enhances DNA-cellulose binding of the [^{3}H]-TA-receptor complexes does not appear to co-purify with the glucocorticoid receptor even through the initial affinity chromatography step. This is suggested by the fact that heating of the partially purified [^{3}H]-TA-receptor complexes in the affinity gel eluate results in a minimal increase (1.5–3.0-fold) in activation as quantitated by DNA-cellulose binding, while heating in a reconstituted system in the presence of unfractionated cytosol facilitates maximal activation (9–10-fold increase when compared to the unactivated control which was not heated). The apparent separation of receptors from the active cytoplasmic "stimulator" at the initial purification step is also confirmed by the fact that the affinity gel supernatant (cytosol remaining after receptors have been adsorbed to affinity resin) is as effective as crude cytosol in stimulating activation in a reconstituted system. The cytoplasmic factor(s) in either crude cytosol or the affinity gel supernatant also appears to stimulate the activation of partially purified [^{3}H]-TA-receptor complexes in a dose-dependent manner. Although we originally reported that the cytoplasmic "stimulator" was stable when incubated at 60°C for 30 min (*33*), we have subsequently demonstrated that it is also stable after heating at 90°C for 30 min. This heat-stable factor is clearly a macromolecule since it is excluded from Sephadex G-25. We have also demonstrated that this factor is sensitive to trypsin and hence is a heat-stable protein.

These preliminary reconstitution data suggested that a heat-stable cytoplasmic protein stimulates activation of partially or highly purified [^{3}H]-TA-receptor complexes based on their ability to bind to DNA-cellulose. We subsequently wanted to demonstrate that this stimulation would be reflected in the expected shift in the elution profile of the activated [^{3}H]-TA-receptor complexes on DEAE-cellulose (*78*). When the purified [^{3}H]-TA-receptor complexes were thermally activated in the presence of the heat-stable protein, approximately a 2-fold increase in DNA-cellulose binding was detected when compared to the heated control. However, despite this 2-fold difference in DNA-cellulose binding, we failed to detect any difference in the percentage of [^{3}H]-TA-receptor complexes eluted in the typical activated position. This somewhat paradoxical result led us to suspect that the heat-stable cytoplasmic protein may not in fact stimulate the conformational change associated with activation which is reflected in the shift in the elution profile of the [^{3}H]-TA-receptor complexes from DEAE-cellulose, but rather may stimulate or enhance the binding of the thermally activated (low salt eluting) complexes to DNA-cellulose. Appropriate experiments were therefore conducted to test this two-step model. Purified [^{3}H]-TA-receptor complexes were thermally activated at 25°C for 30 min. Following this first incubation (step 1) the sample was adjusted to 10 mM MoO_4 to block any further activation (shift to low salt eluting form on DEAE-cellulose). Also, at the conclusion of step 1 the DNA-cellulose binding of the purified [^{3}H]-TA-receptor complexes was assayed and activation also quantitated by DEAE-cellulose chromatography. During step 2 the thermally activated complexes were incubated for an additional 30 min at 25°C in the presence or absence of the cytoplasmic "stimulator." The data generated from experiments such as the one described are consistent with the proposed model. They suggest that during the first molybdate-sensitive step, elevated temperature results in the appropriate shift in the elution profile of purified [^{3}H]-TA-receptor complexes on DEAE-cellulose but only in a 1.5 to 3.0-fold increase in the binding of these complexes to DNA-cellulose. During the second molybdate-insensitive step, a heat-stable cytoplasmic protein interacts with these thermally activated [^{3}H]-TA-receptor complexes and enhances their binding to DNA-cellulose. In crude cytosol these two steps would presumably occur simultaneously and addition of molybdate prior to heating would block the first step and hence the substrate ([^{3}H]-TA-receptor complexes eluting at low salt from DEAE-cellulose) for the second step would not be generated.

Recently we have attempted to identify the active cytoplasmic "stimulator." Several lines of evidence, which will not be presently discussed, indicate that this "stimulator" is not a heat-stable protease or phosphatase. This "stimulator" is clearly distinct from other heat-stable activities in rat liver cytosol which have been reported to affect glucocorticoid receptors (*35*, *51*). The stability of the "stimulator" at elevated temperature (90°C for 30 min) suggests that this endogenous protein may be ribonuclease (RNase), an extremely well characterized enzyme which is known to be heat stable (*21*). We have therefore tested the effects of exogenous pancreatic RNase A on the DNA-cellulose binding of thermally activated, purified [^{3}H]-TA-receptor complexes and have found that it mimics the effects of the heat-stable cytoplasmic "stimulator." Additionally, the 90°C-treated cytosol contains significant RNase activity and when analyzed by SDS-polyacrylamide gel electrophoresis contains a major protein of the same molecular weight as exogenous RNase A. We have therefore tentatively identified the heat-stable "stimulator" as RNase.

If endogenous cytosolic RNase is the heat-stable "stimulator" which facilitates maximal activation in a reconstituted system, then it seems reasonable to predict that inhibition of this endogenous enzyme should block activation of [^{3}H]-TA-receptor complexes present in unfractionated cytosol. To test this hypothesis we have investigated the effects of RNasin, which is a protein (Mr=50,000) purified from the human placenta which specifically inhibits (noncompetitively) RNase activity by binding noncovalently to the enzyme (*15*). Preliminary experiments demonstrate that addition of RNasin (like molybdate) prior to thermal activation significantly blocks the increase in DNA-cellulose binding detected after thermal activation. These data are thus consistent with the concept of endogenous cytoplasmic RNase playing a crucial role in the *in vitro* activation of [^{3}H]-TA-receptor complexes. In this regard it is important to mention that at least seven different RNA degradative enzymes have been described in various subcellular fractions prepared from rat liver tissue (*77*). However, the only degradative enzyme which is heat-stable and hence similar to the cytoplasmic "stimulator" which we have described is a supernatant or mitochondrial alkaline RNase which has a pH optimum between pH 6.7 and 8.5 (*12*).

This possible association of RNA with the purified, unactivated [^{3}H]-TA-receptor complexes and the potential role of endogenous RNase in enhancing the DNA-cellulose binding of the activated complexes raise questions concerning the potential role of RNA in steroid hormone action. Numerous published reports demonstrate that RNA may play a specific role in the nuclear retention of various steroid hormone receptors (*10*, *53*). Likewise, it has been reported that the interaction of a variety of steroid-receptor complexes with DNA-cellulose is inhibited by RNA and that this inhibition is specific with respect to the base composition of the RNA (*25*, *54*, *93*). Several investigators have also reported that exogenous RNase affects unpurified, activated glucocorticoid receptors (*16*, *44*, *45*, *76*). Recently Tymoczko and Phillips (*92*) have reported that both RNase A and RNase T_1 can stimulate the binding of activated rat hepatic dexamethasone-receptor complexes to DNA-cellulose and also shift the sedimentation profile from the 7–8S form to the 3–4S form. The ability of MoO_4, when added prior to heating at 20°C for 15 min, to block the RNase mediated increase in the DNA-cellulose binding in those published experiments is consistent with our hypothesis that MoO_4 blocks the first step in activation. The possible physiological role of an RNA species associated with the unactivated glucocorticoid-receptor complexes is unclear. Our data suggest that this association is not artifactual and that this RNA must be hydrolyzed, presumably by endogenous RNase, before the activated [^{3}H]-TA-receptor complexes can bind maximally to DNA-cellulose. We are currently investigating the possibility that this represents a unique species of RNA which is inaccessible to endogenous RNase in the unactivated glucocorticoid-receptor complexes but becomes accessible only after the first step of activation has been completed.

IV. PHOSPHORYLATION OF THE GLUCOCORTICOID RECEPTOR *IN VIVO*

In order to label the glucocorticoid receptor *in vivo*, adrenalectomized male rats were each injected with 10 mCi [^{32}P]-orthophosphate and 18 hr later their livers were perfused and hepatic cytosol was prepared. The [^{3}H]-TA-receptor complexes were then purified by the three-step procedure (*33*) which has already been described. ^{32}P-containing material was found to co-elute with the peaks of [^{3}H]-TA from both the Bio Gel A-1.5 m column and the subsequent DEAE-cellulose column. Aliquots of the receptor peak fractions eluted from DEAE-cellulose were subsequently subjected to SDS-polyacrylamide gel electrophoresis and staining with Coomassie Blue. As expected, the stained gel again displayed a single Mr=90,000 band in the receptor-peak fractions and the intensity of this band paralleled the distribution of bound [^{3}H]-TA. The autoradiogram of this gel also displayed a pattern of the ^{32}P-labeled Mr= 90,000 protein identical to the Coomassie Blue staining pattern. These data thus demonstrated that the glucocorticoid receptor (Mr=90,000) incorporates ^{32}P *in vivo* and they confirmed the results obtained earlier with mouse L cell receptors (*43*). Singh *et al.* (*88*), who have incubated a rat liver mince with [^{32}P]-orthophosphate, have also confirmed that the hepatic glucocorticoid receptor is a phosphoprotein.

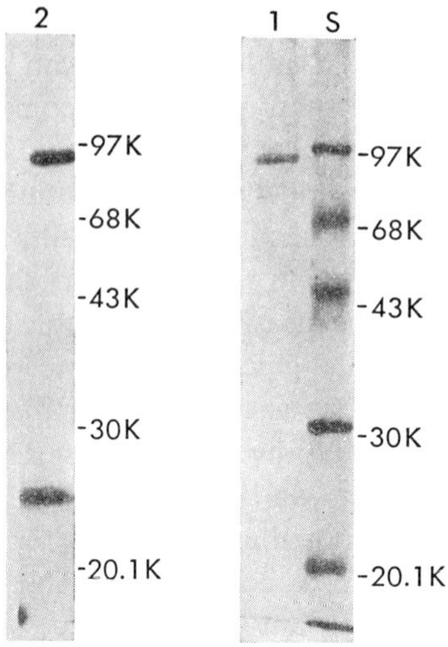

Fig. 4. SDS-polyacrylamide gel electrophoresis of ^{32}P-labeled receptors of the pooled DEAE-fractions. S, standard proteins; lane 1, Coomassie Blue stained final receptor; lane 2, autoradiograph of lane 1. Reproduced with permission from *Biochem. Biophys. Res. Commun.* (*32*).

When the [^{3}H]-TA-receptor complexes eluted from DEAE-cellulose were pooled and subsequently electrophoresed and autoradiographed, another ^{32}P component, which was undetected on the Coomassie Blue stained gel, was visualized (Fig. 4). This phosphorylated component was subsequently shown to be primarily eluted from DEAE-cellulose at a salt concentration much higher than that required for the elution of the purified, unactivated [^{3}H]-TA-receptor complexes. This band also appeared to be identical to the Mr= 24,000 component which we had previously shown to be related to the purified [^{3}H]-TA-receptor complex and detectable with the more sensitive double staining procedure (*33*). Based on protein staining this Mr=24,000 component appeared to be much more heavily phosphorylated than the Mr=90,000 receptor protein itself, and we speculated that it may actually represent a polynucleotide which is tightly associated with the purified, unactivated [^{3}H]-TA-receptor complexes. In light of the more recent data, which has already been discussed, which suggest that an RNA species may be associated with the unactivated complexes, this speculation appears warranted. We have not, however, repeated these *in vivo* phosphorylation experiments and tested the effects of the endogenous heat-stable cytoplasmic "stimulator," which we believe is RNase, on the appearance of this phosphorylated Mr=24,000 component. However, preliminary experiments (M. Webb, unpublished data) demonstrate that incubation of purified unactivated [^{3}H]-TA-receptor complexes with exogenous RNase results in the disappearance of this Mr=24,000 component from SDS gels subjected to the sensitive double staining procedure. In a related study Housley and Pratt (*42*) have demonstrated that when mouse L cells are incubated in the presence of [^{32}P]-orthophosphate, several ^{32}P-labeled components including the 92K receptor protein itself, a doublet at 37K, a heavily phosphorylated species at 21K, and a band at 12K are detected in the final DEAE-cellulose eluate. However, they found that treatment with exogenous RNase eliminated the ^{32}P detected at the 21K position. This 21K phosphorylated component, which may be identical to the 24K phosphorylated component which we have described, may thus be an RNA species which co-purifies with the unactivated [^{3}H]-TA-receptor complexes from these cultured mouse fibroblasts. With regard to the phosphorylation data generated in our laboratory, it is important to note that the Mr=41,000 and 40,000 proteins, which we have previously shown to also be related to the highly purified, unactivated [^{3}H]-TA-receptor complexes (*33*), are not phosphorylated *in vivo*. If these two components represent proteolytic fragments generated during purification of the unactivated receptor protein (Mr=90,000) then it seems reasonable to expect that they would also contain some ^{32}P-labeled residues. Although these two components have not been further characterized, it presently appears rather unlikely that they are generated artifactually from the intact Mr=90,000 receptor protein. Finally, in preliminary experiments we have thermally activated the purified, unactivated [^{3}H]-TA-receptor complexes which had incorporated ^{32}P *in vivo*. The activated complexes were then visualized by DEAE-cellulose chromatography and appeared to contain both ^{3}H and ^{32}P. These preliminary unpublished data suggest that if receptor dephosphorylation occurs during activation then it clearly does not involve complete dephosphorylation of the Mr=90,000 receptor protein itself. Obviously, however, they do not exclude the possibility that activation involves dephosphorylation of a limited number of residues.

V. PROTEIN KINASE ACTIVITY ASSOCIATED WITH PURIFIED GLUCOCORTICOID RECEPTORS

Subsequent to demonstrating that the glucocorticoid receptor is phosphorylated *in vivo* (*32*), and intrigued by the preliminary report that the purified subunits of the chicken progesterone receptor demonstrate protein kinase activity (*26*), we initiated a series of experiments to determine if the purified glucocorticoid receptor is capable of mediating autophosphorylation. In these experiments the purified [^{3}H]-TA-receptor complexes eluted from DEAE-cellulose were incubated with [γ-^{32}P]-ATP for 30 min at 25°C and then the various samples were electrophoresed and the gels were stained, dried, and subjected to autoradiography. The preliminary data which were generated clearly demonstrated that protein kinase activity is associated with the purified rat hepatic glucocorticoid receptor (*62*). These data also demonstrated that incorporation of ^{32}P into the purified Mr=90,000 receptor (and not into the Mr=24,000 component which is phosphorylated *in vivo*) was Ca^{2+} dependent and that Mg^{2+} could not substitute for Ca^{2+} in the kinase reaction. Although we have no explanation for the discrepancy, Kurl and Jacob (*49*) reported that in their hands the kinase activity associated with the purified rat hepatic glucocorticoid receptor is Mg^{2+}-dependent. As seen in Table III, in our experiments neither exogenous cyclic nucleotides (cAMP or cGMP) or exogenous calmodulin promoted phosphorylation in the absence of Ca^{2+}. Also, addition of the chelator EGTA in the presence of Ca^{2+} blocked the phosphorylation reaction. Histones (calf thymus type III) appeared to be phosphorylated when incubated with the purified [^{3}H]-TA-receptor complexes and [γ-^{32}P]-ATP in the absence of Ca^{2+}, but when Ca^{2+} was present the Mr=90,000 receptor protein was preferentially phosphorylated. Finally, these preliminary experiments suggested that the Ca^{2+}-dependent autophosphorylation of the purified receptor does not occur in the absence of bound ligand.

We have now analyzed this autophosphorylation in much greater detail and in so doing have generated new information concerning the preferential ability of the activated receptor complexes to display kinase activity (A. Miller-Diener, T. Schmidt, and G. Litwack, manuscript submitted). Our preliminary observation that the autophosphorylation of the purified [^{3}H]-TA-receptor complexes was Ca^{2+} dependent prompted us to investigate the potential role of calmodulin more thoroughly. As previously mentioned, ex-

TABLE III
Summary of the Effects of Exogenous Reagents on Kinase Activity Associated with Purified [^{3}H]-TA-Receptor Complexes

Reagents tested	Effect on kinase activity
Reagents tested individually	
Ca^{2+}	Stimulation
Mg^{2+}	No effect
cAMP	No effect
cGMP	No effect
Calmodulin	No effect
Reagents tested in presence of Ca^{2+}	
EGTA	Inhibition
Molybdate	Inhibition
Trifluoperazine	Inhibition
Chlorpromazine	Inhibition
Calmidazolium	No effect
Calmodulin	No effect
Fluorosulfonylbenzoyl adenosine	Inhibition
8-Azido ATP	Inhibition

ogenous calmodulin (in the absence of Ca^{2+}) failed to stimulate kinase activity. Subsequently we noted that exogenous calmodulin did not potentiate the autophosphorylation even in the presence of Ca^{2+}. These data suggested that this phosphorylation is not a calmodulin-dependent reaction. However, the possibility existed that enough endogenous calmodulin co-purified with the [^{3}H]-TA-receptor complexes to support maximal autophosphorylation when exogenous Ca^{2+} was added. It should be noted that even when the gels were stained with Coomassie Blue and silver, endogenous calmodulin was not detected, although exogenous calmodulin was easily detected using this sensitive staining procedure. In order to rule out the potential involvement of endogenous calmodulin a number of inhibitors were tested. As seen in Table III, both trifluoperazine (TFP) and chlorpromazine (CPZ), which are relatively nonspecific calmodulin inhibitors, inhibited phosphorylation in the presence of Ca^{2+}. In contrast, calmidazolium, which is a very specific and potent inhibitor of Ca^{2+}-calmodulin dependent enzymes (*94*), did not affect autophosphorylation of the purified receptor. This latter result clearly demonstrates that the Ca^{2+}-dependent kinase activity associated with the purified [^{3}H]-TA-receptor complexes is not calmodulin dependent. Phenothiazines such as TFP and CPZ are known to intercalate in the lipid-rich portions of membranes and nonspecifically inhibit membrane bound protein kinases and receptors (*94*). It is therefore possible that a lipid is associated with the purified receptor which might account for the effects of these nonspecific inhibitors on the kinase activity. Several published studies (*3*, *95*) indicate that Ca^{2+} may interact directly with the glucocorticoid receptor. For example, Van Bohemen *et al.* (*95*) have demonstrated that Ca^{2+} decreases the affinity of crude receptors for synthetic glucocorticoids such as dexamethasone and that this effect cannot be blocked by calmidazolium. In contrast, TFP competitively inhibits the binding of dexamethasone, not *via* calmodulin inhibition, but apparently through a direct interaction with the unpurified glucocorticoid receptor (*96*).

The dependency of the kinase activity associated with the purified [^{3}H]-TA-receptor complexes on an ATP-binding site has also been demonstrated. Both fluorosulfonylbenzoyl 5′-adenosine and 8-azido-ATP, ATP analogues which inhibit a number of protein kinases (*39*, *40*, *105*), completely block phosphorylation of the purified receptors incubated with [γ-^{32}P]-ATP and Ca^{2+} (Table III). Although these data demonstrate the requirement for an ATP binding site, they obviously do not identify the moiety which contains that site. Future experiments involving affinity labeling with fluorosulfonyl [^{14}C]-benzoyl 5′-adenosine or photoaffinity labeling with 8-azido [α-^{32}P]-ATP should facilitate identification of the ATP-binding site (hence kinase) as an integral part of the purified (Mr=90,000) glucocorticoid receptor.

As previously mentioned, the Ca^{2+}-dependent autophosphorylation of the purified receptor appears to require bound ligand. When the [^{3}H]-TA bound to the purified receptors was stripped by treatment with dextran-coated charcoal, ^{32}P-labeling of the receptor protein was no longer detected. If the purified, unbound receptors were reincubated with [^{3}H]-TA for several hr, binding of [^{3}H]-TA and autophosphorylation were again detected. If the stripped, unactivated receptors were incubated with glucocorticoid antagonists such as [^{3}H]-progesterone or [^{3}H]-cortexolone, or with β-lapachone, a nonsteroidal compound which interacts competitively with the ligand binding site (*82*), no subsequent kinase activity was detected despite the fact that these ligands did rebind to the purified receptor. Thus it appeared that only ligands which are glucocorticoid agonists and hence facilitate activation of the receptor (*11*, *61*) will mediate or potentiate the kinase reaction. Numerous experiments also demonstrated that if 10 mM MoO_4, which we had previously shown inhibits activation of purified receptor complexes (*33*), was added prior to the kinase reaction, no autophosphorylation was detected. What is important to note is that the kinase incubation is performed at 25°C for 30 min, which are conditions which will result in considerable thermal activation of the purified [^{3}H]-TA-receptor complexes (*33*). Taken collectively, these data suggested to us that perhaps only the activated complexes demonstrate kinase activity. The results from several subsequent experiments have in fact supported this theory.

If purified, unactivated complexes were first thermally activated at 25°C for 30 min and then incubated with Ca^{2+} and [γ-^{32}P]-ATP in the presence of molybdate, significant incorporation of ^{32}P into the receptor was detected. In the unactivated control, which was also incubated with Ca^{2+} and [γ-^{32}P]-ATP in the presence of molybdate, no incorporation of ^{32}P was detected. The inability of the purified receptors which were rebound with [^{3}H]-progesterone, [^{3}H]-cortexolone, or β-lapachone to demonstrate kinase activity in the earlier experiments probably reflects the fact that these ligands did not facilitate activation during the kinase incubation. Although the data thus support a model in which the purified activated, but not the unactivated, glucocorticoid-receptor complexes incorporate ^{32}P, they do not conclusively identify the kinase molecule itself. One possibility is that the Mr=90,000 receptor protein itself is a kinase and mediates its own phosphorylation. A second possibility is that a distinct Ca^{2+}-dependent kinase co-purifies with the unactivated [^{3}H]-TA-receptor complexes from DEAE-cellulose and, although undetected using a sensitive double staining procedure, is present during the kinase incubation. Knowing that only the activated complexes demonstrate kinase activity, we have performed the following experiment to address this second possibility. Purified, unactivated [^{3}H]-TA-receptor complexes were first thermally activated (25°C for 30 min) and then rechromatographed on DEAE-cellulose. The low salt-eluting, activated [^{3}H]-TA-receptor complexes were then incubated with Ca^{2+} and [γ-^{32}P]-ATP and significant phosphorylation of the Mr= 90,000 receptor protein was detected. We feel that it is highly unlikely that the elution pattern of a contaminating kinase would shift to the exact same position as the activated [^{3}H]-TA-receptor complexes unless that kinase is an integral part of the receptor. In contrast to the situation with the purified subunits of the progesterone receptor (*100*), DEAE-cellulose chromatography does not appear to facilitate separation of a contaminating kinase from the highly purified glucocorticoid receptor. Additional experiments are currently being planned to generate proof that the purified, activated receptor is in fact a Ca^{2+}-dependent kinase.

Finally, recent data have indicated that during this kinase reaction primarily threonine and some serine residues in the receptor protein become phosphorylated. Based on a rough calculation of the ^{32}P:^{3}H ratio of the purified receptors, a maximum of one to five sites appear to be phosphorylated per receptor molecule. The precise function of these phosphorylated residues and the kinase reaction itself is obviously unknown, despite the fact that phosphorylation/dephosphorylation have been speculated to regulate numerous receptor functions. Since histones can be phosphorylated by the activated glucocorticoid receptor it is premature, but tempting, to speculate that receptor mediated phosphorylation of nuclear proteins may play a role in gene activation by glucocorticoids. In this same vein it is conceivable that dephosphorylation of the receptor may "turn off" its interaction with specific genomic acceptor sites. Subsequent rephosphorylation, perhaps mediated by the receptor itself, may also play a role in the recycling of the receptor from the nucleus to the cytoplasm where it then rebinds another glucocorticoid molecule.

SUMMARY

A three-step scheme (affinity chromatography, gel filtration, DEAE-cellulose chromatography) for the purification (approximately 4,000-fold) of unactivated rat hepatic glucocorticoid receptors has been developed in our laboratory. The purified receptor displays hydrodynamic properties, characteristic of molybdate-stabilized, crude hepatic glucocorticoid-receptor complexes. The purified, unactivated glucocorticoid receptors can be partially thermally activated as evidenced by their increased affinity for DNA-cellulose. However, maximal activation can only be achieved in a reconstituted system which includes a heat-stable cytoplasmic protein or "stimulator." Exogenous pancreatic RNase mimics the stimulatory effects of this heat-stable protein on purified receptors and RNasin, a specific RNase inhibitor, blocks thermal activation of crude glucocorticoid-receptor complexes. Taken collectively, these and other data are consistent with a two step model for activation. During the first molybdate-sensitive step, elevated temperature results in the appropriate shift in the elution profile of purified, activated glucocorticoid-receptor complexes on DEAE-cellulose but only in a minimal increase in the binding of these complexes to DNA-cellulose. During the second molybdate-insensitive step, a heat-stable cytoplasmic protein, which we have tentatively identified as endogenous RNase, enhances the binding of these thermally activated complexes to DNA-cellulose.

Additional data demonstrate that unactivated glucocorticoid-receptor complexes are phosphorylated *in vivo* following injection of [^{32}P]-orthophosphate into adrenalectomized rats. Recent experiments also suggest that protein kinase activity is associated with the purified glucocorticoid-receptor complexes. When the thermally activated, but not unactivated, purified receptor complexes are subsequently incubated in the presence of molybdate at 25°C for 30 min with [γ-^{32}P]-ATP and Ca^{2+}, the Mr=90,000 receptor protein becomes phosphorylated. This kinase activity is not dependent on cyclic nucleotides (cAMP or cGMP) or calmodulin. This phosphorylation reaction can be blocked by the addition of sodium molybdate or by stripping the ligand from the purified unactivated receptors, both of which block subsequent thermal activation during the kinase assay. Although there is a remote possibility that a contaminating kinase co-purifies with the glucocorticoid-receptor complexes, additional data strongly suggest that the activated rat hepatic Mr=90,000 receptor protein, in addition to containing the ligand binding site, also functions as a Ca^{2+}-dependent kinase which mediates autophosphorylation.

Acknowledgments

First, and most importantly, I would like to express my sincere appreciation to Drs. Yukiya Sakamoto and Fumihide Isohashi for their kind invitation to participate in this symposium on Mechanisms of Glucocorticoid Hormone Action at the Institute for Cancer Research, Osaka University Medical School. In this regard I would also like to congratulate the other invited participants on their excellent presentations. Secondly, I would like to express my gratitude and appreciation to those colleagues, particularly Drs. P. Grandics, M. Webb, and Mrs. A. Miller-Diener, whose research efforts have contributed significantly to the experiments discussed in my lecture and in this chapter. Thirdly, I wish to acknowledge the technical assistance of Mr. Melvin Paden and Mr. Vincent Pearson of the Fels Research Institute and Mr. Timothy Volm and Mr. Michael Puk of The University of Iowa. Finally, I would like to thank Mrs. Carol Johnson for her excellent typing of this chapter. The research discussed in this chapter was supported at the Temple University School of Medicine by Research Grants AM 13531 and AM 32870 from the National Institute of Arthritis, Metabolism, and Digestive Diseases, NIH;

by PCM 8215844 from the National Science Foundation; by BC 361B from the American Cancer Society and by Core Grant CA 12227 from the National Cancer Institute to the Fels Research Institute. At The University of Iowa this work was supported by Institutional Research Grant IN-122E from the American Cancer Society to the University of Iowa; by a Helmar Maage Memorial Grant for Cancer Research from the American Cancer Society (BC 464); and by research grant 1 R01 Am 34490-01 PC from the National Institute of Arthritis, Metabolism, and Digestive Diseases, NIH. T. Schmidt is a Scholar of the Leukemia Society of America, Inc.

REFERENCES

1) Andrews, P. (1964) Estimation of the molecular weights of proteins by Sephadex gel-filtration. *Biochem. J.* **91**, 222–233.
2) Andrews, P. (1965) The gel-filtration behavior of proteins related to their molecular weights over a wide range. *Biochem. J.* **96**, 595–606.
3) Arányi, P. and Náray, A. (1980) Effect of calcium on chick thymus glucocorticoid receptor. *J. Steroid Biochem.* **12**, 267–272.
4) Axén, R., Porath, J., and Ernbäck, S. (1967) Chemical coupling of peptides and proteins to polysaccharides by means of cyanogen halides. *Nature* **214**, 1302–1304.
5) Bailly, A., Sallas, N., and Milgrom, E. (1977) A low molecular weight inhibitor of steroid receptor activation. *J. Biol. Chem.* **252**, 858–863.
6) Bailly, A., Savouret, J.-F., Sallas, N., and Milgrom, E. (1978) Factors modifying the equilibrium between activated and non-activated forms of steroid-receptor complexes. *Eur. J. Biochem.* **88**, 623–632.
7) Barnett, C.A., Palmour, R.M., Litwack, G., and Seegmiller, J.E. (1983) *In vitro* stabilization of the unoccupied glucocorticoid receptor by adenosine-5′-phosphate. *Endocrinology* **112**, 2059–2068.
8) Barnett, C.A., Schmidt, T.J., and Litwack, G. (1980) Effects of calf intestinal alkaline phosphatase, phosphatase inhibitors and phosphorylated compounds on the rate of activation of glucocorticoid-receptor complexes. *Biochemistry* **19**, 5446–5455.
9) Barnett, C.A. and Wicks, W.D. (1971) Regulation of phosphoenolpyruvate carboxy kinase and tyrosine transaminase in hepatoma cell cultures. I. Effects of glucocorticoids, N6, O2′-dibutryl cycline-cyclic adenosine 3:5′-monophosphate and insulin in Reuber H35 cells. *J. Biol. Chem.* **246**, 7201–7206.
10) Barrack, E.R. and Coffey, D.S. (1980) The specific binding of estrogens and

androgens to the nuclear matrix of sex hormone responsive tissues. *J. Biol. Chem.* **255**, 7265–7275.
11) Baxter, J.D., Rousseau, G.G., Benson, M.C., Garcea, R.L., Ito, J., and Tomkins, G.M. (1972) Role of DNA and specific cytoplasm receptors in glucocorticoid action. *Proc. Natl. Acad. Sci. U.S.A.* **69**, 1892–1896.
12) Beard, J.R. and Razzell, W.E. (1964) Puirfication of alkaline ribonuclease II from mitochondrial and soluble fractions of liver. *J. Biol. Chem.* **239**, 4186–4193.
13) Beato, M., Kalimi, M., and Feigelson, P. (1973) Interaction of glucocorticoids with rat liver nuclei. II. Studies on the nature of the cytosol transfer factor and the nuclear acceptor site. *Biochemistry* **12**, 3372–3379.
14) Bell, P.A. and Munck, A. (1973) Steroid-binding properties and stabilization of cytoplasmic glucocorticoid receptors from rat thymus cells. *Biochem. J.* **136**, 97–107.
15) Blackburn, P., Wilson, G., and Moore, S. (1977) Ribonuclease inhibitor from human placenta. Purification and properties. *J. Biol. Chem.* **252**, 5904–5910.
16) Chong, M.T. and Lippman, M.E. (1982) Effects of RNA and ribonuclease on the binding of estrogen and glucocorticoid receptors from MCF-7 cells to DNA-cellulose. *J. Biol. Chem.* **257**, 2996–3002.
17) Cidlowski, J.A. and Richon, V. (1983) Physical characterization and purification of affinity labeled human glucocorticoid receptors. In "Proceedings of the 64th Annual Meeting of the Endocrine Society," abstract, p. 588.
18) Claman, H.N. (1972) Glucocorticoids and lymphoid cells. *New Engl. J. Med.* **287**, 388-397.
19) DiSorbo, D.M., Phelps, D.S., and Litwack, G. (1980) Chemical probes of amino acid residues affect the active sites of the glucocorticoid receptor. *Endocrinology* **106**, 922–929.
20) Dougherty, J.J., Puri, R.K., and Toft, D.O. (1982) Phosphorylation *in vitro* of chicken oviduct progesterone receptor. *J. Biol. Chem.* **257**, 14226–14230.
21) Dubos, R.J. and Thompson, R.H.S. (1938) Decomposition of yeast nucleic acid by a heat-resistant enzyme. *J. Biol. Chem.* **124**, 501–510.
22) Eisen, H.J., Schleenbaker, R.E., and Simons, S.S., Jr. (1981) Affinity labeling of the rat liver glucocorticoid receptor with dexamethasone 21-mesylate. *J. Biol. Chem.* **256**, 12920–12925.
23) Erdos, T., Best-Belpomme, M., and Bessada, R. (1970) A rapid assay for binding estradiol to uterine receptors. *Anal. Biochem.* **37**, 244–252.
24) Failla, D., Tomkins, G.M., and Santi, D.V. (1975) Partial purification of a glucocorticoid receptor. *Proc. Natl. Acad. Sci. U.S.A.* **72**, 3849–3852.
25) Feldman, M., Kallos, J., and Hollander, V.P. (1981) RNA inhibits estrogen receptor binding to DNA. *J. Biol. Chem.* **256**, 1145–1148.

26) Garcia, T., Touhimaa, P., Mester, J., Buchou, T., Renoir, J.-M., and Baulieu, E.-E. (1983) Protein kinase activity of purified components of the chicken oviduct progesterone receptor. *Biochem. Biophys. Res. Commun.* **113**, 960–966.
27) Goidl, J.A., Cake, M.H., Dolan, K.P., Parchman, L.G., and Litwack, G. (1977) Activation of the rat liver glucocorticoid-receptor complex. *Biochemistry* **6**, 2125–2130.
28) Govindan, M.V. and Manz, B. (1980) Three-step purification of glucocorticoid receptors from rat liver. *Eur. J. Biochem.* **108**, 47–53.
29) Govindan, M.V., Speiss, E., and Majors, J. (1982) Purified glucocorticoid-hormone complex from rat liver cytosol binds specifically to cloned mouse mammary virus long terminal repeats *in vitro*. *Proc. Natl. Acad. Sci. U.S.A.* **79**, 5157–5161.
30) Grandics, P. (1981) Preparation of polysaccharide derivatives containing amino groups for application in affinity chromatography. *Biochem. Biophys. Acad. Sci. Hung.* **16**, 51–55.
31) Grandics, P., Gasser, D.L., and Litwack, G. (1982) Monoclonal antibodies to the glucocorticoid receptor. *Endocrinology* **111**, 1731–1733.
32) Grandics, P., Miller, A., Schmidt, T.J., and Litwack, G. (1984) Phosphorylation *in vivo* of rat hepatic glucocorticoid receptor. *Biochem. Biophys. Res. Commun.* **120**, 59–65.
33) Grandics, P., Miller, A., Schmidt, T.J., Mittman, D., and Litwack, G. (1984) Purification of the unactivated glucocorticoid receptor and its subsequent *in vitro* activation. *J. Biol. Chem.* **259**, 3173–3180.
34) Grandics, P., Puri, R.K., and Toft, D.O. (1982) A new affinity resin for purification of non-transformed avian progesterone receptor. *Endocrinology* **110**, 1055–1057.
35) Grippo, J.F., Tienrungroj, W., Dahmer, M.K., Housley, P.R., and Pratt, W.B. (1983) Evidence that the endogenous heat-stable glucocorticoid receptor-activating factor is thioredoxin. *J. Biol. Chem.* **258**, 13658–13664.
36) Grody, W.W., Schrader, W.T., and O'Malley, B.W. (1982) Activation, transformation, and subunit structure of steroid hormone receptors. *Endocr. Rev.* **3**, 141–163.
37) Harmon, J.M., Eisen, H.J., Brower, S.T., Simons, S.S., Jr., Langley, C.L., and Thompson, E.B. (1984) Identification of human leukemic glucocorticoid receptors using affinity labeling and anti-human glucocorticoid receptor antibodies. *Cancer Res.* **44**, 4540–4547.
38) Harmon, J.M., Schmidt, T.J., and Thompson, E.B. (1984) Molybdate-sensitive and molybdate-resistant activation—labile glucocorticoid-receptor mutants of the human lymphoid cell line CEM-C7. *J. Steroid Biochem.* **21**, 227–236.

39) Hathaway, G.M., Zoller, M.J., and Traugh, J.A. (1981) Identification of the catalytic subunit of casein kinase II by affinity labeling with 5′-*p*-fluorosulfonylbenzoyl adenosine. *J. Biol. Chem.* **256**, 11442–11446.
40) Hoppe, J., and Friest W. (1979) Localization of the high-affinity ATP site in adenosine-3′:5′-monophosphate-dependent protein kinase type I. Photoaffinity labeling studies with 8-azido adenosine 5′-triphosphate. *Eur. J. Biochem.* **93**, 141–146.
41) Housley, P.R., Dahmer, M.K., and Pratt, W.B. (1982) Inactivation of glucocorticoid-binding capacity by protein phosphatases in the presence of molybdate and complete reactivation by dithiothreitol. *J. Biol. Chem.* **257**, 8615–8618.
42) Housley, P.R. and Pratt, W.B. (1984) Small [^{32}P]-labeled components of L-cells that co-elute with the molybdate-stabilized glucocorticoid receptor during purification. *Fed. Proc.* **43**, 908 (abstract).
43) Housley, P.R. and Pratt, W.B. (1983) Direct demonstration of glucocorticoid receptor phosphorylation by intact L-cells. *J. Biol. Chem.* **258**, 4630–4635.
44) Hutchens, T.W., Markland, F.S., and Hawkins, E.F. (1982) Glucocorticoid receptor activation is a reversible process. In "Proceedings of the 64th Annual Meeting of the Endocrine Society," abstract, p. 568.
45) Hutchens, T.W., Markland, F.S., and Hawkins, E.F. (1982) RNA induced reversal of glucocorticoid receptor activation. *Biochem. Biophys. Res. Commun.* **105**, 20–27.
46) Irie, S., Sezaki, M., and Kato, Y. (1982) A faithful double stain for proteins in the polyacrylamide gels with Coomassie Blue and silver. *Anal. Biochem.* **126**, 350–354.
47) John, J. and Moudgil, V. (1979) Activation of glucocorticoid receptor by ATP. *Biochem. Biophys. Res. Commun.* **90**, 1242–1248.
48) Kalimi, M., Colman, P., and Feigelson, P. (1975) The "activated" hepatic glucocorticoid-receptor complex. Its generation and properties. *J. Biol. Chem.* **250**, 1080–1086.
49) Kurl, R.N., and Jacob, S.T. (1984) Phosphorylation of purified glucocorticoid receptor from rat liver by an endogenous protein kinase. *Biochem. Biophys. Res. Commun.* **119**, 700–705.
50) Leach, K.L., Dahmer, M.K., Hammond, N.D., Sando, J.J., and Pratt, W.B. (1979) Molybdate inhibition of glucocorticoid receptor inactivation and transformation. *J. Biol. Chem.* **254**, 11884–11890.
51) Leach, K.L., Grippo, J.F., Housley, P.R., Dahmer, M.K., Salive, M.E., and Pratt, W.B. (1982) Characterization of an endogenous glucocorticoid receptor stabilizing factor. *J. Biol. Chem.* **257**, 381–388.
52) LeFevre, B., Bailly, A., Sallas, N., and Milgrom, E. (1979) Activated steroid-

receptor complex. Comparison of assays using DNA-cellulose or homologous nuclei. *Biochim. Biophys. Acta* **585**, 266–272.

53) Liang, T. and Liao, S. (1974) Association of the uterine 17β-estradiol-receptor complex with ribonucleoprotein *in vitro* and *in vivo*. *J. Biol. Chem.* **249**, 4671–4678.

54) Liao, S., Smythe, S., Tymoczko, J.L., Rossini, G.P., Chen, C., and Hiipakka, R.A. (1980) RNA-dependent release of androgen and other steroid-receptor complexes from DNA. *J. Biol. Chem.* **255**, 5545–5551.

55) Lustenberg, P., Formstecher, P., and Dautrevaux, M. (1981) Purification of rat liver glucocorticoid receptor by affinity chromatography: design of a suitable adsorbent. *J. Steroid Biochem.* **14**, 697–703.

56) March, S.C., Parikh, I., and Cuatrecasas, P. (1974) A simplified method for cyanogen bromide activation of agarose for affinity chromatography. *Anal. Biochem.* **60**, 149–152.

57) Markovic, R.D. and Litwack, G. (1980) Activation of liver and kidney glucocorticoid-receptor complexes occurs *in vivo*. *Arch. Biochem. Biophys.* **202**, 374–379.

58) Migliaccio, A., Lastoria, S., Moncharmont, B., Rotondi, A., and Auricchio, F. (1982) Phosphorylation of calf uterus 17β-estradiol receptor by endogenous Ca^{2+} stimulated kinase activating the hormone binding of the receptor. *Biochem. Biophys. Res. Commun.* **109**, 1002–1010.

59) Migliaccio, A., Rotondi, A., and Auricchio, F. (1984) Calmodulin-stimulated phosphorylation of 17β-estradiol receptor on tyrosine. *Proc. Natl. Acad. Sci. U.S.A.* **81**, 5921–5925.

60) Milgrom, E. (1981) Activation of steroid-receptor complexes. In "Biochemical Actions of Hormones," Vol. 13, ed. Litwack, G., pp. 465–492, Academic Press, New York.

61) Milgrom, E., Atger, M., and Baulieu, E.-E. (1973) Acidophilic activation of steroid hormone receptors. *Biochemistry* **12**, 5198–5205.

62) Miller, A.S., Schmidt, T.J., and Litwack, G. (1984) Protein kinase activity of purified rat hepatic glucocorticoid receptor. In "Excerpta Medica, Abstracts of the 7th International Congress of Endocrinology," abstract, p. 1056, Elsevier Science Publishers Co., Inc., New York.

63) Miller, L.K., Tuazon, F.B., Niu, E.-M., and Sherman, M.R. (1981) Human breast tumor estrogen receptor: effects of molybdate and electrophoretic analyses. *Endocrinology* **108**, 1369–1378.

64) Miyabe, S. and Harrison, R.W. (1983) *In vivo* activation and nuclear binding of the AtT-20 mouse pituitary tumor cell glucocorticoid receptor. *Endocrinology* **112**, 2174–2180.

65) Munck, A. and Brinck-Johnsen, T. (1968) Specific and non-specific physiochem-

ical interactions of glucocorticoids and related steroids with rat thymus cells *in vitro*. *J. Biol. Chem.* **243**, 5556–5565.

66) Munck, A. and Foley, R. (1979) Activation of steroid hormone-receptor complexes in intact target cells in physiological conditions. *Nature* **278**, 752–754.

67) Munck, A. and Leung, K. (1981) Glucocorticoid receptors and mechanisms of action. In "Receptors and Mechanisms of Action of Steroid Hormones," ed. Pasqualini, J.R., pp. 311–397, Dekker, New York.

68) Murphy, R.F., Conlon, J.M., Inman, A., and Kelly, G.J.C. (1977) Comparison of non-biospecific effects in immunoaffinity chromatography using cyanogen bromide and bifunctional oxirane as immobilizing agents. *J. Chromatogr.* **135**, 427–433.

69) Nielsen, C.J., Sando, J.J., and Pratt, W.B. (1977) Evidence that dephosphorylation inactivates glucocorticoid receptors. *Proc. Natl. Acad. Sci. U.S.A.* **74**, 1398–1402.

70) Norman, M.R. and Thompson, E.B. (1977) Characterization of a glucocorticoid-sensitive human lymphoid cell line. *Cancer Res.* **37**, 3785–3791.

71) O'Farrell, P.H. (1975) High resolution two-dimensional electrophoresis of proteins. *J. Biol. Chem.* **250**, 4007–4021.

72) Parchman, L.G. and Litwack, G. (1977) Resolution of activated and unactivated forms of the glucocorticoid receptor from rat liver. *Arch. Biochem. Biophys.* **183**, 374–382.

73) Parikh, I., March, S.C., and Cuatrecasas, P. (1974) Topics in the methodology of substitution reactions with agarose. In "Methods in Enzymology," eds., Jakoby, W.B. and Wilchek, N., Vol. 34, pp. 77–102, Academic Press, New York.

74) Robins, D.M., Paek, I., Seeburg, P.H., and Axel, R. (1982) Regulated expression of human growth hormone genes in mouse cells. *Cell* **29**, 623–631.

75) Rosen, J.M., Rosen, F., Milholland, R.J., and Nichol, C.A. (1970) Effects of cortisol on DNA metabolism in the sensitive and resistant lines of mouse lymphoma. *Cancer Res.* **30**, 1129–1136.

76) Rossini, G.P. and Barbiroli, B. (1983) RNase-sensitive glucocorticoid-receptor complexes form HELA cell nuclei. *Biochem. Biophys. Res. Commun.* **113**, 876–882.

77) Roth, J.S. (1967) Some observations on the assay and properties of ribonucleases in normal and tumor tissues. In "Methods in Cancer Research," Vol. 111, ed. Busch, H., pp. 153–242, Academic Press, New York.

78) Sakaue, U. and Thompson, E.B. (1977) Characteristics of two forms of glucocorticoid hormone-receptor complex separated by DEAE-cellulose column chromatography. *Biochem. Biophys. Res. Commun.* **77**, 533–541.

79) Sando, J.J., Hammond, N.D., Stratford, C.A., and Pratt, W.B. (1979) Activation of thymocyte glucocorticoid receptors to the steroid binding form. The roles of reducing agents, ATP, and heat-stable factors. *J. Biol. Chem.* **254**, 4779–4789.
80) Sando, J.J., LaForest, A.C., and Pratt, W.B. (1979) ATP-dependent activation of L cell glucocorticoid receptors to the steroid binding form. *J. Biol. Chem.* **254**, 4772–4778.
81) Scheidereit, C., Geisse, S., Westphal, H.M., and Beato, M. (1983) The glucocorticoid receptor binds to defined nucleotide sequences near the promoter of mouse mammary tumor virus. *Nature* **304**, 749–752.
82) Schmidt, T.J., Miller-Diener, A., and Litwack, G. (1984) β-Lapachone, a specific competitive inhibitor of ligand binding to the glucocorticoid receptor. *J. Biol. Chem.* **259**, 9536–9543.
83) Schmidt, T.J. and Litwack, G. (1982) Activation of the glucocorticoid-receptor complex. *Physiol. Rev.* **62**, 1131–1192.
84) Schmidt, T.J., Sekula, B.C., and Litwack, G. (1981) The effects of 1, 10-phenanthroline of the binding of activated rat hepatic glucocorticoid receptors to deoxyribonucleic acid-cellulose. *Endocrinology* **109**, 803–812.
85) Sekula, B.C., Schmidt T.J., and Litwack, G. (1981) Redefinition of modulator as an inhibitor of glucocorticoid receptor activation. *J. Steroid Biochem.* **14**, 161–166.
86) Sherman, M.R., Moran, M.C., Neal, R.M., Niu, E.-M., and Tuazon, F.B. (1982) Characterization of molybdate-stabilized glucocorticoid receptors in healthy and malignant tissues. In "Progress in Research and Clinical Applications of Corticosteroids," eds. Lee, H.J. and Fitzgerald, T.J., pp. 45–66, Heyden and Sons, Philadelphia.
87) Simons, S.S., Jr. and Thompson, E.B. (1981) Dexamethasone 21-mesylate: an affinity label of glucocorticoid receptors from rat hepatoma tissue culture cells. *Proc. Natl. Acad. Sci. U.S.A.* **78**, 3541–3545.
88) Singh, V., Eessalu, T., Ghag S., and Moudgil, V. (1984) Phosphorylation of rat liver glucocorticoid receptor. In "Excerpta Medica, Abstracts of the 7th International Congress of Endocrinology," abstract, p. 2095, Elsevier Science Publishers Co., Inc., New York.
89) Sundberg L. and Porath, J. (1974) Preparation of adsorbents for biospecific affinity chromatography. Attachment of group-containing ligands to insoluble polymers by means of bifunctional oxiranes. *J. Chromatogr.* **90**, 87–98.
90) Thompson, E.B. and Lippman, M.E. (1974) Mechanism of action of glucocorticoids. *Metabolism* **23**, 159–202.
91) Tomkins, G.M., Gelehrter, T.D., Granner, D.K., Martin D.W., Samuels,

H.S., and Thompson, E.B. (1969) Control of specific gene expression in higher organisms. *Science* **166**, 1474–1480.

92) Tymoczko, J.L. and Phillips, M.M. (1983) The effects of ribonuclease on rat liver dexamethasone receptor: increased affinity for deoxyribonucleic acid and altered sedimentation profile. *Endocrinology* **112**, 142–149.

93) Tymoczko, J.L., Shapiro, J., Simenstad, D.J., and Nish, A.D. (1982) The effects of polyribonucleotides on the binding of dexamethasone-receptor complex to DNA. *J. Steroid Biochem.* **16**, 595–598.

94) Van Belle, H. (1981) R 24571: A potent inhibitor of calmodulin-activated enzymes. *Cell Calcium* **2**, 483–494.

95) Van Bohemen, C.G., Eliard, P.H., and Rousseau, G.G. (1983) Control by calcium ions of steroid binding to the glucocorticoid receptor. *Mol. Physiol.* **4**, 95–110.

96) Van Bohemen, C.G. and Rousseau, G.G. (1982) Calmodulin antagonists competitively inhibit dexamethasone binding to the glucocorticoid receptor. *FEBS Lett.* **143**, 21–25.

97) Vedeckis, W.V. (1983) Subunit dissociation as a possible mechanism of glycocorticoid receptor activation. *Biochemistry* **22**, 1983–1989.

98) Weatherill, P.J. and Bell, P.A. (1982) Characterization of the molybdate-stabilized glucocorticoid receptor from rat thymus. *Biochem. J.* **206**, 633–640.

99) Webb, M., Miller-Diener, A.S., and Litwack, G. (1985) Purification characterization, and activation of the glucocorticoid-receptor complex from rat kidney cortex. *Biochemistry*, in press.

100) Weigel, N.L. (1984) Isolation of protein kinases from chicken oviduct which phosphorylate the progesterone receptor *in vitro*. In "Excerpta Medica, Abstracts of the 7th International Congress of Endocrinology," abstract, p. 2710, Elsevier Science Publishers Co., Inc., New York.

101) Weigel, N.L., Tash, J.S., Means, A.R., Schrader, W.T., and O'Malley, B.W. (1981) Phosphorylation of hen progesterone receptor by cAMP dependent protein kinase. *Biochem. Biophys. Res. Commun.* **102**, 513–519.

102) Wheeler, R.H., Leach, K.L., LaForest, A.C., O'Toole, T.E., Wagner, R., and Pratt, W.B. (1981) Glucocorticoid receptor activation and inactivation in cultured human lymphocytes. *J. Biol. Chem.* **256**, 434–441.

103) Wilchek, M., Oka, T., and Topper, Y.J. (1975) Structure of a soluble superactive insulin is revealed by the nature of the complex between cyanogen bromide activated sepharose and amines, *Proc. Natl. Acad. Sci. U.S.A.* **72**, 1055–1058.

104) Wray, W., Boulikas, T., Wray, V.P., and Hancock, R. (1981) Silver staining of proteins in polyacrylamide gels. *Anal. Biochem.* **118**, 197–203.

105) Zoller, M.J. and Taylor, S.S. (1979) Affinity labeling of the nucleotide binding site of the catalytic subunit of cAMP-dependent protein kinase using *p*-fluorosulfonyl-[^{14}C] benzoyl 5′-adenosine. *J. Biol. Chem.* **254**, 8363–8368.

3

Glucocorticoid Hormone: Mechanisms of Action, ed. Sakamoto, Y. and Isohashi, F., pp. 69–94, Japan Sci. Soc. Press, Tokyo/Springer-Verlag, Berlin (1986)

TRANSLOCATION OR BINDING OF ACTIVATED RECEPTOR-GLUCOCORTICOID COMPLEX TO NUCLEAR ACCEPTORS: IDENTIFICATION AND MECHANISM OF POSSIBLE REGULATORY SUBSTANCES

FUMIHIDE ISOHASHI and YUKIYA SAKAMOTO

Department of Biochemistry, Institute for Cancer Research, Osaka University Medical School, Osaka 530, Japan

During the last ten years a great deal of information about the molecular mechanisms of the actions of steroid hormones has been accumulated. This has provided a better understanding of these mechanisms and indicated how complicated they are. Elucidation of this problem is very difficult because there are very low concentrations of specific steroid receptors in target cells and these are labile during ordinary biochemical manipulations. However, it is becoming apparent that specific cytoplasmic receptors mediate modulation of gene expression by steroid hormones by which the hormones exert specific effects in target tissues. To cause the modulation of gene expression in the nuclei of target cells, a steroid hormone binds to a specific cytoplasmic receptor and then becomes activated or transformed to a form that can bind to nuclei, chromatin, and DNA, and be translocated and bind to nuclear components (for reviews, see refs. *20, 24, 25, 60, 61, 69*). Therefore, for convenience in analysis of the molecular mechanism of steroid action *in vitro*, the underlying physiological process is divided into three steps: first, a steroid

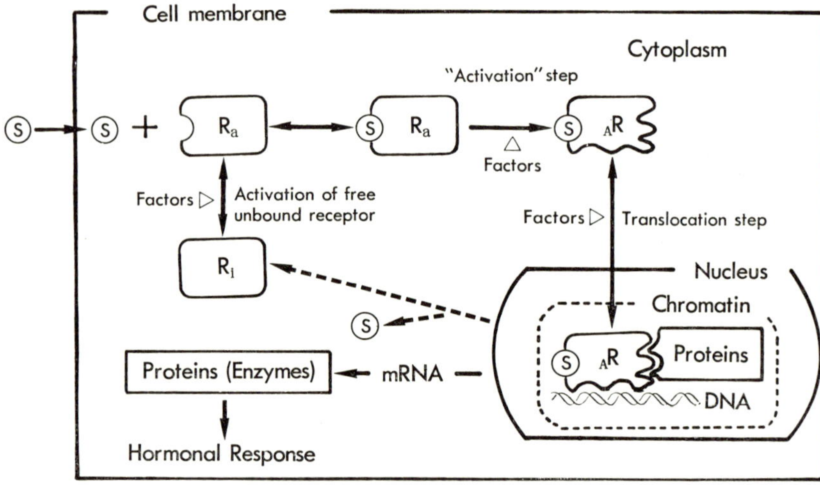

Fig. 1. Schematic representation of the steps in glucocorticoid action. S, steroid; R_i, aporeceptor; R_a, holoreceptor; $_AR$, activated receptor-steroid complex. The direct action of the steroid-receptor complex is not illustrated other than that *via* nuclear chromatin, but this does not mean that it is excluded from the mechanisms.

binding step, which involves binding of the steroid hormone to a cytoplasmic holoreceptor derived from an aporeceptor by a regulatory mechanism (*53*, *54*, *62*, *63*), as shown in Fig. 1; second, an activation or transformation step (for review, see ref. *67*), which involves conversion of the steroid-receptor complex to an activated receptor-steroid complex, which is a form able to bind to nuclei, chromatin, and DNA (Fig. 1); third, a translocation or binding step, which involves migration of the activated or transformed receptor-steroid complex to the nuclei and its binding to chromosomal components (for reviews, see refs. *20*, *73*) (Fig. 1). This review summarizes some of our studies on the receptor "translocation" or "binding" of the already activated receptor-glucocorticoid complex to nuclear components. Some possible molecular mechanisms are discussed based on our results and those of others.

I. BACKGROUND INFORMATION ON THE TRANSLOCATION STEP AND MODULATORS

At present, little is known about the properties and mechanism of the translocation step and its regulatory factors. However, the existence of a macromolecular receptor translocation inhibitor(s) was reported as early as 1974 in relation to the problem of the pseudo-saturability of nuclear acceptor sites in a cell-free estrogen-receptor system (*7*). Pseudo-saturation was also observed in the rat liver glucocorticoid-receptor system due to the inhibition by a non-receptor macromolecular receptor translocation inhibitor(s) by its interaction with the "activated" receptor-glucocorticoid complex (*2*, *3*, *35*, *72–74*). Thus the macromolecular receptor translocation inhibitor(s) has been studied only incidentally. The significance of these macromolecules in the cytosol is enigmatic and these molecules are often considered to be nonspecific contaminants of the preparations formed during processing of tissues. Namely, they may be artifacts formed by modification of the receptor rather than being an integral part of the regulatory mechanism of steroid hormone action. These historical developments in study of translocation modulators have been described in a previous review (*21*).

However, Liu and Webb (*32*) presented evidence that a cytoplasmic macromolecular receptor translocation inhibitor does in fact limit the amount of receptor-glucocorticoid entering the nucleus of intact Novikoff hepatoma cells, which are relatively glucocorticoid resistant. Moreover, the association of "activated" receptor-steroid with nuclei *in vivo* seems to be a regulated process, because accumulation of activated receptor-glucocorticoid complex in the cytoplasm in a time-dependent manner after glucocorticoid treatment has been observed in rat thymus cells (*49*), rat liver (*34*), and kidney (*34*), indicating the physiological significance of the activation step (*34*, *49*).

In addition to macromolecular factors, some endogenous and exogenous materials were also reported to inhibit the binding of already activated receptor-steroid complex. These materials were pyridoxal-5′-phosphate (*5*, *12*, *30*, *46*, *47*, *55*, *71*), rifamycin AF/013 (*33*, *66*, *76*), *o*-phenanthroline (*33*, *68*, *76*), aurintricarboxylic acid (*37–39*, *45*), polyribonucleotides with certain types of bases (*27–29*), and some metal ions (*50–52*) such as molybdate, tungstate, and vanadete. In contrast, zinc potentiated the binding of androgen receptor to nuclei (*9*).

II. LOW-MOLECULAR WEIGHT TRANSLOCATION MODULATORS

1. Interaction with a Macromolecular Translocation Inhibitor(s)

A crude macromolecular translocation inhibitor(s) (MTI(s)), which was a fraction of the liver cytosol of adrenalectomized rats free from receptor and low-molecular weight material, inhibited the nuclear binding of highly purified (3,000–5,000-fold) activated receptor-glucocorticoid complex, as shown in Fig. 2 (*22*). As we reported previously (*22*), dialyzed receptor-free cytosol (free from low-molecular weight material) was more inhibitory than undialyzed cytosol. This finding suggested the presence of a dialyzable substance(s) in the cytosol that interferes with the inhibition by a nondialyzable macromolecular substance(s) of the binding of activated receptor complexes to nuclei (*22*). Pooled small molecular weight fractions, called low-molecular weight translocation modulator(s) (LTM(s)), eluted from Sephadex G-25 had weaker inhibitory effects than the crude MTI(s) on the binding of the activated receptor complex to nuclei from rat liver (*22*). When the two components were mixed, their inhibitory effects decreased, as shown in Fig. 2

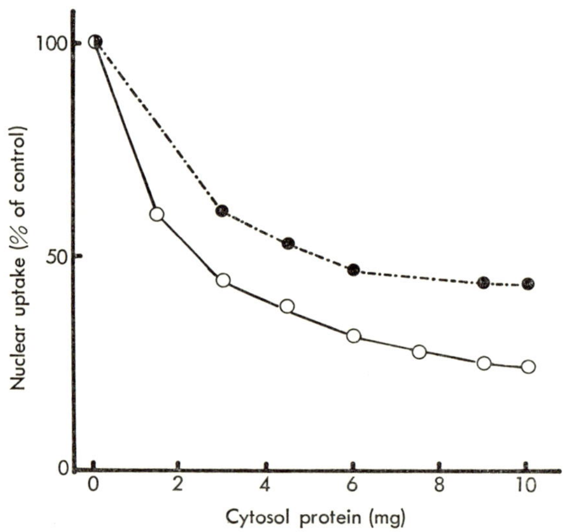

Fig. 2. Inhibitory effects of dialyzed receptor-free cytosol as crude macromolecular translocation inhibitor(s) (MTI(s)) and effects of the low-molecular weight translocation modulators (LTMs) on the inhibitory action of crude MTI(s). Increasing amounts of the crude MTI(s) from the liver cytosol of adrenalectomized (ADX) rats were added to nuclear binding assay medium containing highly purified activated receptor [^{3}H]-triamcinolone acetonide complex and nuclei with (●) or without (○) LTMs separated from the cytosol on Sephadex G-25. Other procedures were as described in ref .*22*.

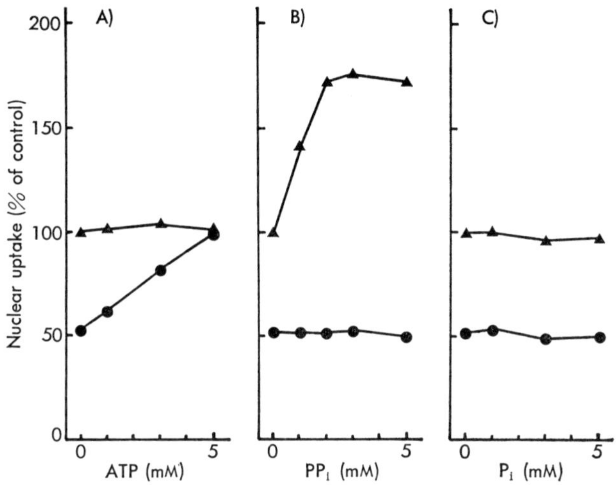

Fig. 3. Effects of ATP (A), pyrophosphate (B), and phosphate (C) on nuclear binding of highly purified activated receptor-glucocorticoid complex with (●) or without (▲) crude MTI(s). Procedures were as described in ref. *13*.

(*22*). This effect was reversible, because removal of the LTM(s) enhanced the inhibitory effect of the crude MTI(s). This LTM(s) thus appeared to differ from reported MTI(s) and low-molecular weight activation inhibitors or factor f, which inhibits the conversion of the "unactivated" receptor-steroid complex to the "activated" form (*4, 6, 11, 26, 31, 67, 70*).

2. *Effects of Nucleotides on the Inhibitory Effect of MTI(s)*

Pyridoxal-5′-phosphate was reported to inhibit the binding of the activated receptor-steroid complex to DNA-cellulose and nuclei (*5, 12, 30, 47, 55*). Addition of pyridoxal phosphate to the crude MTI(s) increased the inhibition additively (*22*), in contrast to LTM(s) which decreased the inhibition (*22*). Next, we tried to purify so-called LTM(s) on a Sephadex G-10 column, and found that the fraction with LTM(s) activity contained materials with UV absorption and molecular weights in the range of those of nucleotides. Thus, we screened nucleotides to find those mimicking the activity of LTM(s) (*15*). Of the various compounds tested at physiological concentrations, only ATP partly mimicked the properties of LTM(s) (*15*). As shown in Fig. 3A, ATP at up to 5 mM specifically decreased the inhibitory effect of the crude MTI(s), but within this concentration range it did not inhibit nuclear binding of highly purified activated complex (*15*). It is noteworthy that dATP (unpublished results), ADP, AMP, cAMP, UTP, TTP, CTP, and GTP had no detectable effects at physiological concentrations (*15*).

It is becoming apparent that an interaction between ATP and steroid receptors may be important in the mechanism of steroid action (*1*, *17*, *18*, *40*, *48*, *53*, *54*, *62*, *63*). ATP has been reported to bind directly to the activated steroid-receptor complex in various steroid receptor systems (*36*, *41–44*). ADP was also effective in a crude receptor system, but to a lesser extent than ATP.

Besides nucleotides, we also tested pyrophosphate (PP_i) and phosphate (P_i), finding that they had no effect on the activity of crude MTI(s) (Fig. 3B and 3C) (*13*). Surprisingly, in the absence of the crude MTI(s), PP_i greatly enhanced the nuclear binding of the highly purified activated receptor-glucocorticoid complex (Fig. 3B) (*13*) as described in more detail in Section IV of this review.

In other studies (*19*), we found that ATP had no appreciable effect on crude MTI(s) prepared from glucocorticoid-resistant tumor cells (AH 130 ascites hepatoma cells), as shown in Fig. 4. The difference in the results (Fig.

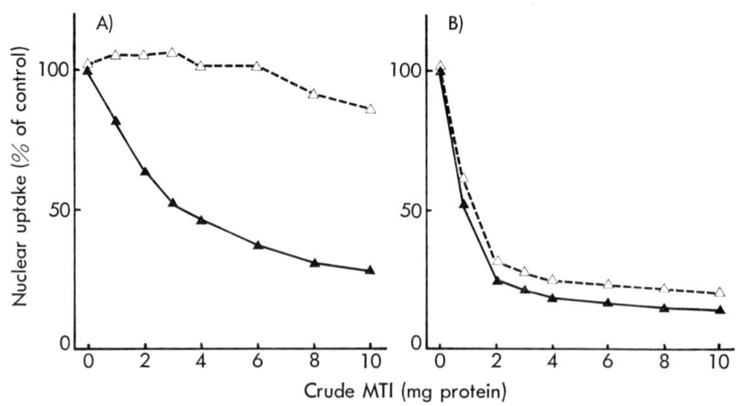

Fig. 4. Effects of 5 mM ATP on the inhibitory activity of various concentrations of crude MTI(s) from the liver of adrenalectomized rats (A) and glucocorticoid-resistant AH 130 tumor cells from adrenalectomized rats (B). Nuclear binding was assayed with (△) or without (▲) 5 mM ATP. Other procedures were as described in ref. *19*.

4) with rat liver and AH 130 cells suggests that these crude MTI(s) preparations consist of different molecules.

III. MACROMOLECULAR TRANSLOCATION MODULATORS

1. Separation of MTI-I, II, and III

We partially purified these inhibitory materials (MTIs) from rat liver cytosol and AH 130 ascites tumor cells by DEAE-cellulose chromatography (*19*, *56*). As shown in Fig. 5, we separated at least three MTIs from both cytosols (*19*, *56*). The first fraction (peak I) (called MTI-I) was eluted with about 0.1 M NaCl, the second (peak II) (called MTI-II) with about 0.2 M NaCl,

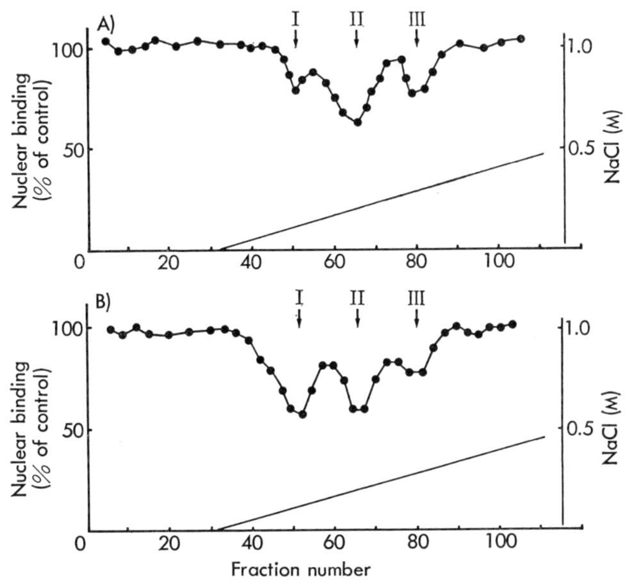

Fig. 5. DEAE-Cellulose column chromatography of MTIs of the cytosol of adrenalectomized rat liver (A), or AH 130 tumor cells (B) harvested from adrenalectomized rats. The material was eluted with a linear gradient of 0–0.5 M NaCl in the buffer used for equilibration of the column. I, peak I; II, peak II; III, peak III. Inhibitory activity was measured by nuclear binding assay with isolated nuclei and highly purified activated receptor complex, both from the liver of adrenalectomized rats. Other procedures were as described in refs. *19* and *56*.

and the third (peak III) (called MTI-III) with about 0.3 M NaCl (*19*, *56*). The positions of elution of the fractions from rat liver and AH 130 cells were similar (*19*). However, unlike in rat liver, peak I (MTI-I) from AH 130 tumor cytosol was larger than peak II (MTI-II) or peak III (MTI-III) (*19*). Moreover, peak I of AH 130 tumor cytosol was much larger than that of rat liver (*19*). Thus, so-called macromolecular receptor translocation inhibitor (non-receptor macromolecule) consists of at least three macromolecules and the inhibitory activity of the crude MTI represents the sum of the effects of the three inhibitors (*19*, *56*). In preliminary studies (unpublished observation) we found by gel filtration that the apparent molecular weights of MTI-I, MTI-II, and MTI-III were about 240,000, 700,000, and 30,000, respectively. But some of the MTIs lost much of their inhibitory activity and tended to aggregate during gel filtration, judging from the considerable tailing and minor peaks associated with the major peaks. Therefore, further studies are needed on these molecular weights. The inhibitory action of the crude MTI(s) was reported to be due to its direct interaction with the activated receptor complex, not with the nuclei, reducing the affinity of the activated receptor complex for the nuclear acceptors (*72*, *74*). But at the moment, it is unknown whether these three inhibitors all associate directly with the receptor. Further purification and identification of the inhibitory macromolecules are in progress.

2. Separation of ATP-stimulated Macromolecular Translocation Promoter (ASTP)

The entities of the inhibitors of the MTIs from rat liver and AH 130 tumor cells were similar, though the proportions of the three fractions from the two sources were quite different (*19*, *56*). These results cannot explain the different ATP-sensitivities of the crude MTI(s) of rat liver (*56*) and AH 130 tumor cells (*19*) shown in Fig. 4. Thus, we next examined the effects of ATP on each of the three peak fractions eluted from a DEAE-cellulose column (*19*, *56*), and results are shown in Fig. 6. In the presence of ATP, peak I from rat liver enhanced nuclear binding of the activated receptor complex (*56*). In contrast, pooled fractions of peak I of AH 130 tumor cell cytosol were not influenced by the addition of ATP (*19*). The inhibitory activity of peak II of either rat liver or AH 130 tumor cells was overcome by 5 mM ATP, whereas the inhibitory effect of peak III from either rat liver or AH 130 tumor cells was little affected by ATP (Fig. 6) (*19*, *56*).

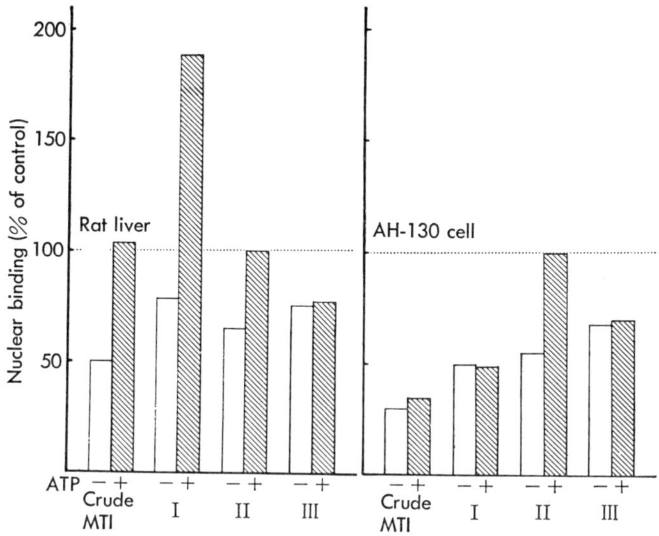

Fig. 6. Effects of ATP on the inhibitory actions of the peak fractions from DEAE-cellulose. I, peak I; II, peak II; III, peak III from DEAE-cellulose. Other procedures were as described in ref. *19*.

The effect of ATP on a combination of the three peak fractions was similar to that on crude MTIs from rat liver and AH 130-cells, suggesting that the apparent effects of ATP observed with the crude systems were results of the combined effects of ATP on ATP-sensitive and -insensitive substances (*56*). However, these results raised the question of whether the ATP-enhancing activity of peak I or the ATP-inhibiting activity of peak II and the inhibitions by peak I and peak II in the absence of ATP are due to the same respective single components. ATP-sensitive material and inhibitory material (MTI-II) in peak II could not separated by DEAE-cellulose column chromatography or gel filtration (*56*). Thus far, we believe that MTI-II itself may be ATP-sensitive (losing inhibitory activity in the presence of ATP). However, the following evidence suggests that the ATP-enhancing activity of peak I and the inhibitory material (MTI-I) are different. First, AH 130 tumor cells were deficient in ATP-enhancing activity in peak I, but contained much MTI-I (*19*, *56*). Second, heat treatment (60°C, 15 min) completely abolished the inhibitory activity of peak I, but enhanced the ATP-enhancing activity (*57*). Third, the ATP-enhancing activity could be separated from the inhibitory

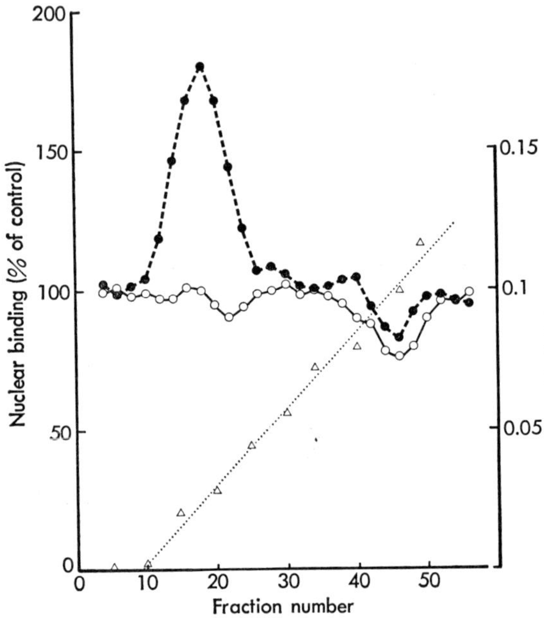

Fig. 7. Elution of the ATP-stimulated factor enhancing nuclear binding and the inhibitor from a DEAE-cellulose column with a moderate linear gradient of NaCl (0–0.13 M). ASTP, ATP-stimulated translocation promoter; I, MTI-I. Nuclear binding assay with (●) or without (○) 5 mM ATP. Other procedures were as described in ref. *57*.

material (MTI-I) by DEAE-cellulose column chromatography on a less steep linear gradient of NaCl (0 to 0.13 M) (Fig. 7) (*57*) than used previously (*19*, *56*). We named this material ATP-stimulated translocation promoter (ASTP). Present information on the macromolecular translocation modulators and their relation to ATP, described above, are shown schematically in Fig. 8.

3. *Purification and Characterization of ASTP*

We purified ASTP from the liver of adrenalectomized rats (*58*). The conventional purification procedure (*58*) involves ammonium sulfate fractionation (23–43 %), DEAE-cellulose column chromatography, hydroxylapatite column chromatography, gel filtration, heat-treatment, and then CM-cellulose column chromatography. The yield is about 10 % from rat liver cytosol, with about 1,000-fold purification (about 20 % pure) (*58*). Further purification

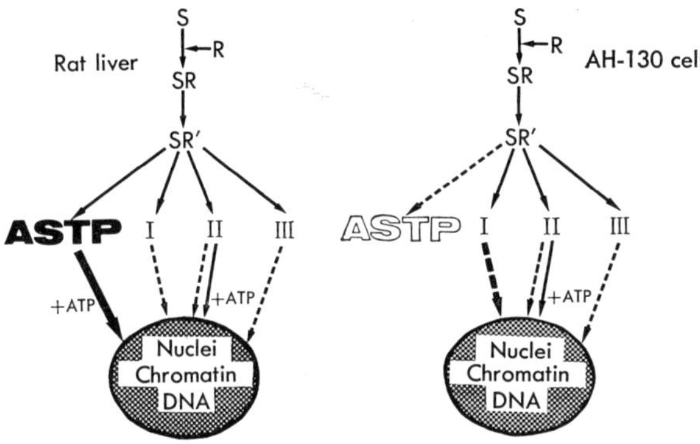

Fig. 8. Schematic representation of the macromolecular translocation modulators. ASTP, which is not present in the cytosol of glucocorticoid resistant AH 130 tumor cells; I, MTI-I, which is predominant in the cytosol of AH 130 tumor cells; II, MTI-II with inhibitory activity that is overcome by ATP; III, MTI-III.

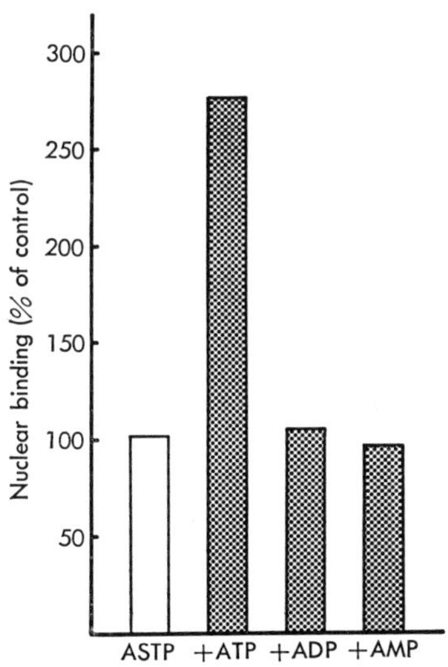

Fig. 9. Effects of ATP, ADP, AMP, on ASTP activity. ASTP alone had no effect on the nuclear binding of the highly purified activated receptor complex. ASTP was purified about 1000-fold.

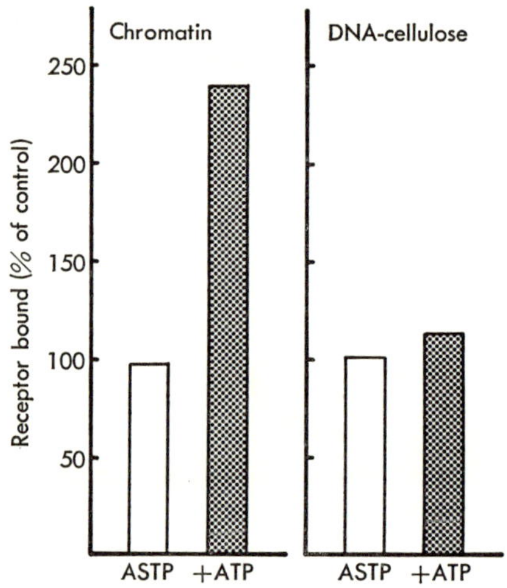

Fig. 10. Effects of ASTP and ATP on chromatin and nuclear binding of the activated receptor-glucocorticoid complex. ASTP was purified about 1000-fold from the liver of adrenalectomized rats. Chromatin was prepared from isolated liver nuclei of adrenalectomized rats. DNA-cellulose, native calf thymus DNA-cellulose.

is in progress. Preliminary results suggest that the ASTP (probably protein) has a molecular weight of about 93,000, and an isoelectric point of about 4.5, loses little activity on heating at 60°C for 5 min, and is RNase resistant. ASTP showed dependency on ATP (*57*), but not on ADP or AMP as seen in Fig. 9 (*58*). It enhanced chromatin binding of the highly purified activated complex in the presence of ATP, but scarcely affected the DNA-cellulose binding (native calf thymus DNA-cellulose) as also shown in Fig. 10 (*58*).

4. Physiological Changes of ASTP Activity

As mentioned earlier, glucocorticoid-resistant AH 130 tumor cells lack ASTP activity (*19*). Our recent preliminary results (*77*) indicated that adrenalectomy of rats significantly enhanced the ASTP activity in the cytosol fraction of their liver, whereas treatment with glucocorticoid decreased the activity (*77*). Furthermore, regenerating liver after partial hepatectomy (about 24 hr after the operation) is known to be relatively glucocorticoid resistant (*23*, *59*) and to show rapid DNA synthesis (*23*, *59*). Interestingly,

ASTP activity was decreased in this period (*77*). Therefore, ASTP activity is changeable under different conditions of the animals (*19*, *77*). An interesting finding was that these changes are roughly parallel to changes of the glucocorticoid receptor levels in the cytosol of rat liver (*23*). From these results, attractive possibilities are that in the presence of ATP, ASTP binds to activated receptor-glucocorticoid complex, increasing the affinity of the latter for nuclear components, but not for DNA, and that the ASTP-receptor complex migrates and is more tightly bound to nuclear acceptors than the activated complex alone. Obviously, more studies are necessary on these possibilities.

IV. REGULATION OF NUCLEAR BINDING OF ACTIVATED RECEPTOR COMPLEX BY PYROPHOSPHATE (PP_i) AND ATP

1. *Enhancement of Nuclear Binding by* PP_i

As mentioned briefly in Section II-2, in the absence of crude MTIs, PP_i greatly

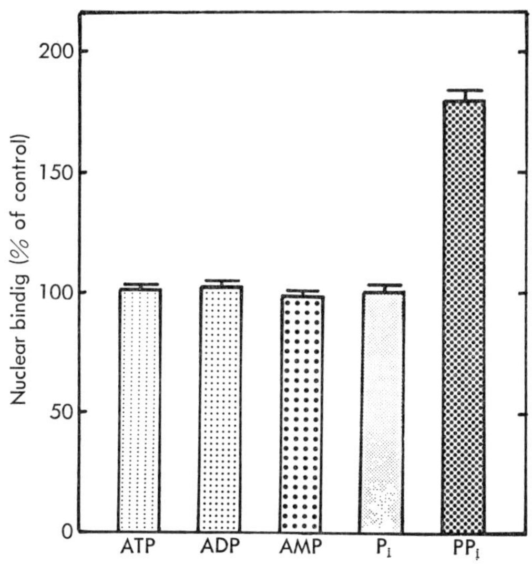

Fig. 11. Effects of phosphorus compounds on nuclear binding of highly purified activated receptor-glucocorticoid complex. Concentrations of 5 mM ATP, ADP, AMP, P_i, or PP_i were added to the mixture for assay of nuclear binding containing 2×10^6 nuclei and highly purified activated receptor-glucocorticoid complex (1.8 pmol). Procedures were as described in refs. *13* and *14*.

enhanced the nuclear binding of highly purified activated receptor complex, while ATP, ADP, AMP, and P_i did not (Fig. 11) (*13*, *14*). However, this effect of PP_i cannot be detected with a crude receptor-nuclear binding system (*13*). This difference may be due to contamination of the cytosol of rat liver with high activity of inorganic pyrophosphatase (*16*), resulting in hydrolysis of added PP_i to phosphate. But purified nuclei of rat liver are known to have little activity of this enzyme (*16*). On the other hand, it is possible that some of the MTIs interfere with the interaction of PP_i with the receptor.

2. Effects of PP_i in the Presence of Nucleotides

Progressive addition of PP_i increased nuclear binding of the highly purified activated complex (*13*, *14*). However, at concentrations of more than 7 mM, PP_i decreased nuclear binding (Fig. 12) (*13*, *14*).

Although ATP affects nuclear binding in the presence of macromolecular translocation modulators (*13*, *15*, *19*, *56*), ATP alone at up to 5 mM did not influence nuclear binding (*13*, *15*). However, in the presence of 5 mM ATP, nuclear binding of the highly purified activated complex decreased

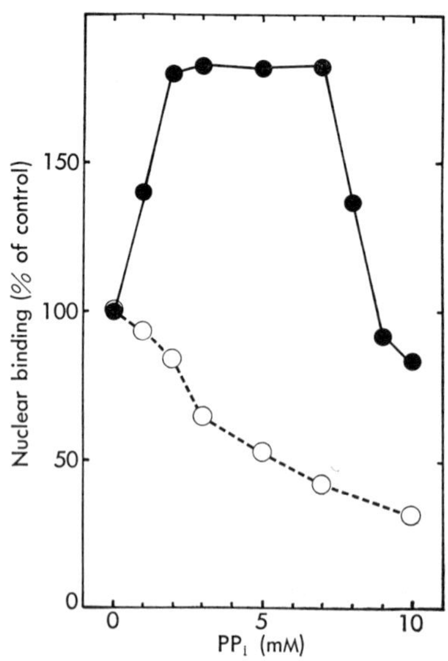

Fig. 12. Effect of the concentration of PP_i with (○) or without (●) 5 mM ATP on nuclear binding of the activated receptor-glucocorticoid complex. Procedures were as described in refs. *13* and *14*.

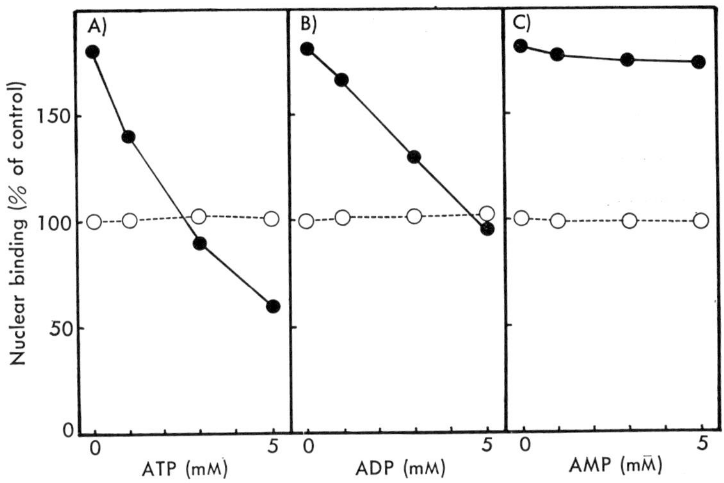

Fig. 13. Effects of ATP (A), ADP (B), and AMP (C) with (●) or without (○) 3 mM PP_i on nuclear binding of highly purified activated receptor-glucocorticoid complex. Procedures were as described in ref. *14*.

steadily with increase in PP_i concentration (*14*). ADP, but not AMP, also decreased the enhancing effect of PP_i, though less than ATP (Fig. 13) (*14*).

3. *Effects of PP_i on Chromatin and DNA-cellulose Binding*

As shown in Fig. 14, ATP had no appreciable effect on chromatin binding (*14*). PP_i greatly enhanced chromatin binding like nuclear binding of the activated complex (*14*). However, under the experimental conditions used, neither ATP nor PP_i caused any appreciable change in DNA-cellulose binding (*14*). These results, together with other results, suggest that nuclear (chromatin) binding and DNA-cellulose binding of activated receptor complex are different; that is, the chromosomal protein binding site(s) and DNA-cellulose binding site(s) on the receptor molecules are different (*14*).

A combination of ATP and PP_i (Fig. 14) decreased nuclear, chromatin, and DNA-cellulose bindings (*14*). These molecular mechanisms will be discussed in a later subsection (IV-5).

Mg^{2+} at 4 mM was necessary for maintenance of the integrity of the nuclei or nuclear membrane in our *in vitro* nuclear binding assay system (*13–15*, *19*, *22*, *56*, *57*). ATP and PP_i are known to be metal chelators. Thus

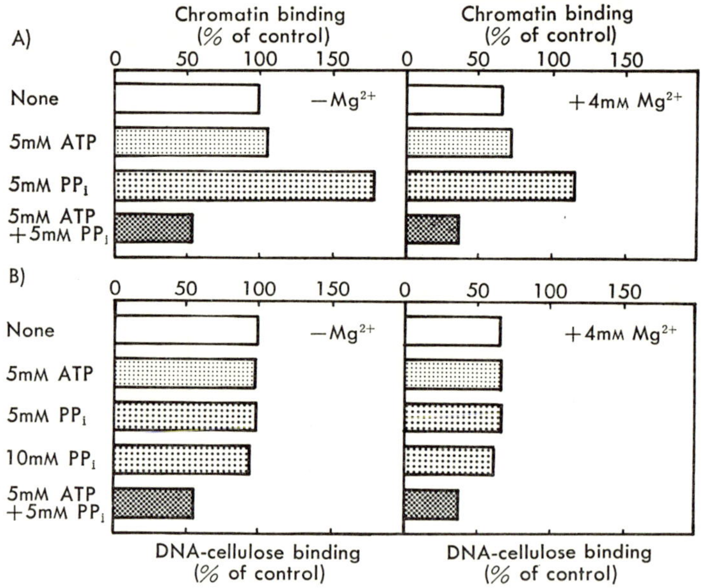

Fig. 14. Effects of ATP, PP_i, and Mg^{2+} on chromatin binding (A) and DNA-cellulose (calf thymus DNA)-binding (B) of highly purified activated receptor-glucocorticoid complex. Values are percentages of the values of the control not containing ATP, PP_i, or $MgCl_2$ for chromatin binding and DNA-cellulose binding. Other procedures were as described in ref. *14*.

we examined whether ATP and PP_i exerted effects as chelators, using chromatin as acceptor in the absence of $MgCl_2$ (Fig. 14) (*14*). In the absence of 4 mM $MgCl_2$, PP_i also enhanced chromatin binding, although chromatin binding without $MgCl_2$ was more than that with 4 mM $MgCl_2$ (*14*). Moreover, 5 mM ATP did not affect chromatin binding, with or without $MgCl_2$, and no difference was observed between the inhibitory effects of ATP with PP_i in the presence and absence of $MgCl_2$ (*14*).

4. ATP and PP_i Binding Site(s) on Activated Receptor-steroid Complex Molecules

The sedimentation coefficient of the highly (3,000–5,000-fold) purified "activated" receptor-glucocorticoid ([^{3}H]-triamcinolone acetonide) complex was about 4.2S and its molecular weight was 98,000 (*14*). Addition of 5 mM ATP

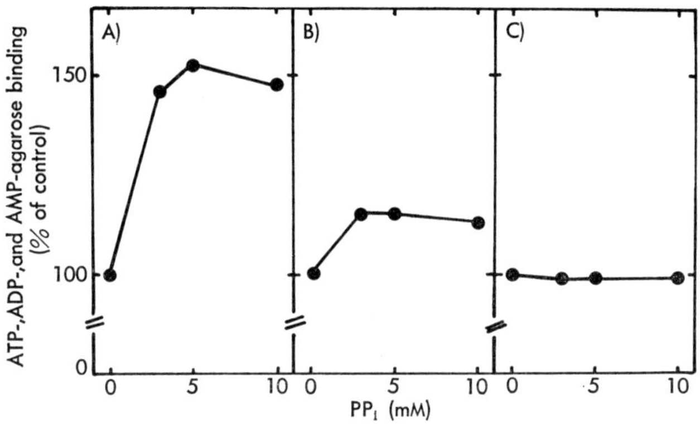

Fig. 15. Effects of PP$_i$ on ATP-(A), ADP-(B), and AMP-agarose (C) binding of highly purified activated receptor-glucocorticoid complex. The control values were 18,000 cpm (ATP-agarose), 2,100 cpm (ADP-agarose), and 2,900 cpm (AMP-agarose). Procedures were as described in ref. *14*.

and/or PP$_i$ to the medium in sucrose density gradient centrifugation and to the medium on gel filtration, both at low salt concentration, did not cause any appreciable change in the sedimentation coefficient or molecular weight of the activated complex (4.2S and 98,000, respectively) (*14*). Thus, ATP and/or PPi did not cause any aggregation of the complexes or their fragmentation.

To obtain more information on the interaction of ATP and/or PP$_i$ with receptor molecules, we tested their effects on the receptor molecules using ATP-agarose (*14*), since the activated receptor complex in various steroid hormone systems is known to bind specifically to an ATP-matrix (*36*, *41*, *42*, *44*). If PP$_i$ binds to the same site(s) as ATP, it should compete with ATP-agarose binding of the "activated" complex. However, as shown in Fig. 15A, PP$_i$ greatly enhanced ATP-agarose binding, suggesting that the binding site(s) for PP$_i$ on the molecule of "activated" receptor complex is distinct from the ATP-binding site(s) (*14*). Similar experiments with ADP-agarose (Fig. 15B) and AMP-agarose (Fig. 15C) demonstrated that PP$_i$ slightly enhanced ADP-agarose binding, but did not affect AMP-agarose binding (*14*). However, the bindings of the receptor to ADP-agarose and AMP-agarose were much lower than that to ATP-agarose (about 12–16 % of the latter) (*14*).

5. Proposed Allosteric Model of the Interaction of ATP and PP_i with the Receptor Molecule and Their Binding Function for Acceptors

The results presented here and reported previously suggest that the mechanism of the effects of ATP and PP_i on receptor binding functions for acceptors is complex. To explain the various observations, we propose the following hypothesis (Fig. 16) (*14*), although it must be pointed out that this scheme is very tentative. Highly (3,000–5,000-fold) purified preparations of the activated receptor(R)-glucocorticoid complex(G) have a molecular weight of about 98,000 and a sedimentation coefficient of 4.2S. They have binding sites for both PP_i and ATP. ATP binding to the activated R-G complex causes no apparent change of the chromatin or DNA binding (DNA-cellulose; DNA-c). However, PP_i binding to the activated R-G complex may cause some allosteric conformational change of the R molecule resulting in a high affinity state of the species for nuclei and chromatin, but not for DNA. Thus the chromatin binding site(s) and DNA-c binding site(s) on the activated R-G complex may be distinct. The concurrent bindings of ATP and PP_i may cause a change of the chromatin and DNA-c binding site(s) on the activated R-G complex, making it less accessible to chromatin and DNA-c. In this way, both PP_i and ATP may regulate the function or the nature of the

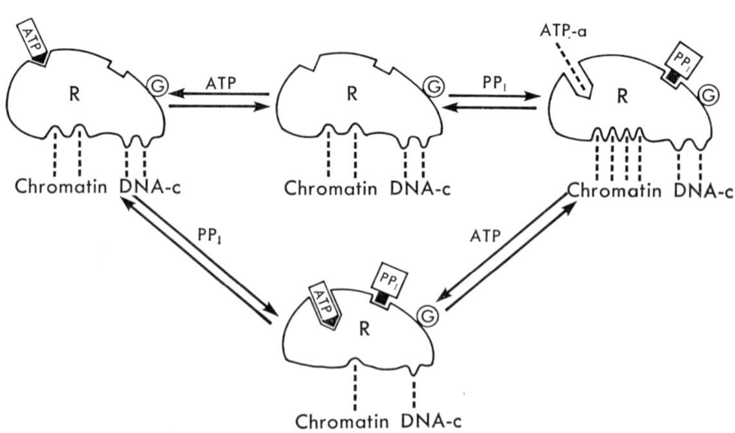

Fig. 16. Proposed scheme for the effects of ATP and PP_i on the chromatin (nuclear) binding site(s) and DNA-cellulose binding site(s). ATP-a, ATP-agarose; G, glucocorticoids; DNA-c, DNA-cellulose.

activated receptor-glucocorticoid complex. Recently, several investigators (*8, 10, 64, 65, 75*) demonstrated that activated receptor-glucocorticoid complex recognizes and binds to regulatory elements of DNA (specific sequences in DNA). However, it is uncertain whether the DNA-c binding site(s) and the specific DNA binding site(s) are related. If they are not, the next question is how ATP and PP_i influence the specific DNA site(s). Further studies with DNA-cellulose using DNA with specific sequences for receptor binding should provide the answer to these questions.

SUMMARY

Studies on the translocation step confirmed the crucial and highly regulated process of translocation or nuclear binding, involving macromolecular and low-molecular weight translocation modulators, in the mechanism of the action of glucocorticoid hormone. The putative MTI in rat liver cytosol consists of at least three inhibitors (MTI-I, MTI-II, and MTI-III). Moreover, results showed that there is at least one ASTP that enhances the nuclear binding of the activated receptor-glucocorticoid complex in the presence of ATP. Besides ASTP, the inhibitory activity of MTI-II was overcome by 5 mM ATP, but ASTP and MTI-II were apparently different entities. Glucocorticoid-resistant AH 130 tumor cells have no ASTP activity, but contain a high concentration of putative MTI, particularly MTI-I. The ASTP activity in the cytosol of rat liver was found to be increased by adrenalectomy and decreased by treatment with glucocorticoids. Thus, these results suggest that some, if not all, of these macromolecular translocation modulators have a significant physiological role. Their mechanisms of action are unknown, but the putative MTI is known to interact directly with the activated receptor complex, reducing the affinity of the latter for nuclear acceptors, such as nuclei, chromatin, and DNA.

LTMs were also found in the cytosol of rat liver. These LTMs caused weak inhibition of the binding of the activated receptor complex to nuclei, but overcame the inhibitory activity of crude MTI. Like LTMs, ATP can overcome the inhibition of crude MTIs. ATP alone caused no inhibition of nuclear binding, but ATP plus PP_i decreased nuclear binding, although PP_i greatly enhanced nuclear binding. These effects are presumed to be due to allosteric change of the acceptor binding sites of the activated receptor complex.

It is concluded from this work that the process of translocation is highly regulated by many endogenous translocation modulators that act by modifying the functions of the activated receptor-steroid complex.

Acknowledgments

Our work reported in this review was supported by a Grant-in-Aid for Cancer Research and a Grant-in-Aid for Scientific Research from the Ministry of Education, Science and Culture of Japan, and grants from Kudo Gakujutsu-Zaidan, the Osaka Anticancer Association, the Osaka Kisoigaku Kenkyu Shoreikai, and the Byotai Taisha Kenkyukai of Japan. The authors thank Drs. M. Horiuchi, K. Okamoto, Y. Mitsui, Y. Nakanishi, T. Matsunaga, K. Tsukanaka, M. Terada, and Messrs. A. Yamamoto, K. Naito, and M. Gotoh for their collaboration, and Miss M. Fukuoka for secretarial assistance.

REFERENCES

1) Andreasen, P.A. (1981) A specific adenine nucleotide effect on the rat liver glucocorticoid receptor, demonstrated by aqueous two-phase partitioning. *Biochim. Biophys. Acta* **676**, 205–212.
2) Atger, M. and Milgrom, E. (1976) Mechanism and kinetics of the thermal activation of glucocorticoid hormone-receptor complex. *J. Biol. Chem.* **251**, 4758–4762.
3) Atger, M. and Milgrom, E. (1978) Interaction of glucocorticoid-receptor complexes with rat liver nuclei. *Biochim. Biophys. Acta* **539**, 41–53.
4) Bailly, A., Sallas, N., and Milgrom, E. (1977) A low molecular weight inhibitor of steroid receptor activation. *J. Biol. Chem.* **252**, 858–863.
5) Cake, M.H., DiSorbo, D.M., and Litwack, G. (1978) Effect of pyridoxal phosphate on the DNA binding site of activated hepatic glucocorticoid receptor *J. Biol. Chem.* **253**, 4886–4891.
6) Cake, M.H., Goidl, J.A., Parchman, L.G., and Litwack, G. (1976) Involvement of a low molecular weight component(s) in the mechanism of action of the glucocorticoid receptor. *Biochem. Biophys. Res. Commun.* **71**, 45–52.
7) Chamness, G.C., Jennings, A.W., and McGuire, W.L. (1974) Estrogen receptor binding to isolated nuclei. A nonsaturable process. *Biochemistry* **13**, 327–331.
8) Chandler, V.L., Maler, B.A., and Yamamoto, K.R. (1983) DNA sequences bound specifically by glucocorticoid receptor *in vitro* render a heterologous promoter hormone responsive *in vivo*. *Cell* **33**, 489–499.

9) Colvard, D.S. and Wilson, E.M. (1984) Zinc potentiation of androgen receptor binding to nuclei *in vitro*. *Biochemistry* **23**, 3471–3478.
10) Geisse, S., Scheidereit, C., Westphal, H.M., Hynes, N.E., Groner, B., and Beato, M. (1982) Glucocorticoid receptors recognize DNA sequences in and around murine mammary tumour virus DNA. *EMBO J.* **1**, 1613–1619.
11) Goidl, J.A., Cake, M.H., Dolan, K.P., Parchman, L.G., and Litwack, G. (1977) Activation of the rat liver glucocorticoid receptor complex. *Biochemistry* **16**, 2125–2130.
12) Hiipakka, R.A. and Liao, S. (1980) Effect of pyridoxal phosphate on the androgen receptor from rat prostate: Inhibition of receptor aggregation and receptor binding to nuclei and to DNA-cellulose. *J. Steroid Biochem.* **13**, 841–846.
13) Horiuchi, M., Isohashi, F., Okamoto, K., Matsunaga, T., Terada, M., Mitsui, Y., and Sakamoto, Y. (1982) Regulation of nuclear uptake of "activated" receptor-glucocorticoid complex by pyrophosphate. *Biochem. Biophys. Res. Commun.* **105**, 146–151.
14) Horiuchi, M., Isohashi, F., Okamoto, K., and Sakamoto, Y. (1984) Partially purified "activated" receptor-glucocorticoid complex from rat liver: Regulation of nuclear, chromatin, and DNA-cellulose binding of "activated" complex by pyrophosphate and ATP. *J. Biochem.* **96**, 727–737.
15) Horiuchi, M., Isohashi, F., Terada, M., Okamoto, K., and Sakamoto, Y. (1981) Interaction of ATP with a macromolecular translocation inhibitor of the nuclear binding of "activated" receptor-glucocorticoid complex. *Biochem. Biophys. Res. Commun.* **98**, 88–94.
16) Irie, M., Yabuta, A., Kimura, K., Shindo, Y., and Tomita, K. (1970) Distribution and properties of alkaline pyrophosphatases of rat liver. *J. Biochem.* **67**, 47–58.
17) Ishii, D.N. and Aronow, L. (1973) *In vitro* degradation and stabilization of the glucocorticoid binding component from mouse fibroblasts. *J. Steroid Biochem.* **4**, 593–603.
18) Ishii, D.N., Pratt, W.B., and Aronow, L. (1972) Steady-state level of the specific glucocorticoid binding component in mouse fibroblasts. *Biochemistry* **11**, 3896–3904.
19) Isohashi, F., Horiuchi, M., Okamoto, K., and Sakamoto, Y. (1984) Separation and characterization of receptor-translocation inhibitors from AH 130 tumor cells. *J. Steroid Biochem.* **20**, 1117–1122.
20) Isohashi, F. and Sakamoto, Y. (1984) Steroid hormone receptors. *Igaku no Ayumi* **129**, 908 (in Japanese).
21) Isohashi, F. and Sakamoto, Y. (1985) Low-molecular-weight and macromolecular translocation modulators affecting the binding of activated receptor-glucocorticoid complex to nuclei, chromatin, and DNA. In "Molecular Mech-

anism of Steroid Hormone Action," ed. Moudgil, V.K., pp. 225-247, Walter de Gruyter & Co., Berlin, New York.

22) Isohashi, F., Terada, M., Tsukanaka, K., Nakanishi, Y., and Sakamoto Y. (1980) A low-molecular-weight translocation modulator and its interaction with a macromolecular translocation inhibitor of the activated receptor-glucocorticoid complex. *J. Biochem.* **88**, 775–781.

23) Isohashi, F., Tsukanaka, K., Terada, M., Nakanishi, Y., Fukushima, H., and Sakamoto, Y. (1979) Alteration in binding of dexamethasone to glucocorticoid receptors in regenerating rat liver after partial hepatectomy. *Cancer Res.* **39**, 5132–5135.

24) Jensen, E.V. and DeSombre, E.R. (1979) Steroid sex hormone receptors and action. In "Endocrinology," eds. DeGroot, L.J., Cahill, G.F., Jr., Odell, W.D., Martini, L., Potts, J.T., Jr., Nelson, D.H., Steinberger, E., and Winegrad, A.I., Vol. 3, pp. 2055–2061, Grune and Stratton, New York.

25) Lan, N.C., Karin, M., Nguyen, T., Weisz, A., Birnbaum, M.J., Eberhardt, N.L., and Baxter, J.D. (1984) Mechanisms of glucocorticoid hormone action. *J. Steroid Biochem.* **20**, 77–88.

26) Leach, K.L., Grippo, J.F., Housley, P.R., Dahmer, M.K., Salive, M.E., and Pratt, W.B. (1982) Characteristics of an endogenous glucocorticoid receptor stabilizing factor. *J. Biol. Chem.* **257**, 381–388.

27) Liang, T. and Liao, S. (1974) Association of the uterine 17β-estradiol-receptor complex with ribonucleoprotein *in vitro* and *in vivo*. *J. Biol. Chem.* **249**, 4671–4678.

28) Liao, S., Hiipakka, R.A., and Judge, S.M. (1984) Steroid receptor mechanism: Androgen action in the rat ventral prostate. In "Biochemical and Biophysical Studies of Proteins and Nucleic Acids," eds. Lo, T.B., Liu, T.Y., and Li, C.H., pp. 123–135, Elsevier Science Publishing Co. Inc., Amsterdam.

29) Liao, S., Smythe, S., Tymoczko, J.L., Rossini, G.P., Chen, C., and Hiipakka, R.A. (1980) RNA-dependent release of androgen- and other steroid-receptor complexes from DNA. *J. Biol. Chem.* **255**, 5545–5551.

30) Litwack, G. and Cake, M.H. (1977) DNA binding site of activated glucocorticoid receptor. Interaction with pyridoxal-P. *Fed. Proc.* **36**, 911a.

31) Litwack, G., Schmidt, T.J., Sekula, B.C., DiSorbo, D.M., Ohl, V.S., Barnett, C.A., and Phelps, D.S. (1982) Activation of the glucocorticoid receptor. *J. Mol. Cell. Cardiol.* **14**, 59–64.

32) Liu, S.H. and Webb, T.E. (1977) Elevated concentration of a dexamethasone-receptor translocation inhibitor in Novikoff hepatoma cells. *Cancer Res.* **37**, 1763–1767.

33) Lohmar, P.H. and Toft, D.O. (1975) Inhibition of the binding of progesterone

receptor to nuclei: Effects of *o*-phenanthroline and rifamycin AF/013. *Biochem. Biophys. Res. Commun.* **67**, 8–15.

34) Marković, R.D. and Litwack, G. (1980) Activation of liver and kidney glucocorticoid-receptor complexes occurs *in vivo*. *Arch. Biochem. Biophys.* **202**, 374–379.

35) Milgrom, E. and Atger, M. (1975) Receptor translocation inhibitor and apparent saturability of the nuclear acceptor. *J. Steroid Biochem.* **6**, 487–492.

36) Miller, J.B. and Toft, D.O. (1978) Requirement for activation in the binding of progesterone receptor to ATP-Sepharose. *Biochemistry* **17**, 173–177.

37) Moudgil, V.K. and Caradonna, V.M. (1982) Inhibition of nuclear uptake of glucocorticoid-receptor complex by aurintricarboxylic acid. *J. Steroid Biochem.* **17**, 585–589.

38) Moudgil, V.K. and Eessalu, T.E. (1980) Sensitivity of progesterone receptors to aurintricarboxylic acid. Inhibition of nuclear uptake, ATP and DNA binding. *Biochim. Biophys. Acta* **627**, 301–312.

39) Moudgil, V.K. and Eessalu, T.E. (1980) Inhibition of rat uterine estradiol receptor by aurintricarboxylic acid. *Life Sci.* **27**, 1159–1167.

40) Moudgil, V.K. and John, J.K. (1980) ATP-dependent activation of glucocorticoid receptor from rat liver cytosol. *Biochem. J.* **190**, 799–808.

41) Moudgil, V.K. and John, J.K. (1980) Interaction of rat liver glucocorticoid receptor with adenosine 5′-triphosphate. *Biochem. J.* **190**, 809–818.

42) Moudgil, V.K. and Toft, D.O. (1975) Binding of ATP to the progesterone receptor. *Proc. Natl. Acad. Sci. U.S.A.* **72**, 901–905.

43) Moudgil, V.K. and Toft, D.O. (1976) ATP-PPi exchange activity of progesterone receptor. *Proc. Natl. Acad. Sci. U.S.A.* **73**, 3443–3447.

44) Moudgil, V.K. and Toft, D.O. (1977) Interaction of progesterone receptor with immobilized adenosine triphosphate. *Biochim. Biophys. Acta* **490**, 477–488.

45) Moudgil, V.K. and Weekes, G.A. (1978) Inhibition of hen oviduct estradiol receptor by aurintricarboxylic acid. *FEBS Lett.* **94**, 324–326.

46) Muldoon, T.G. and Cidlowski, J.A. (1980) Specific modification of rat uterine estrogen receptor by pyridoxal 5′-phosphate. *J. Biol. Chem.* **255**, 3100–3107.

47) Müller, R.E., Traish, A., and Wotiz, H.H. (1980) Effects of pyridoxal 5′-phosphate on uterine estrogen receptor. I. Inhibition of nuclear binding in cell-free system and intact uterus. *J. Biol. Chem.* **255**, 4062–4067.

48) Munck, A. and Brinck-Johnson, T. (1968) Specific and nonspecific physicochemical interactions of glucocorticoids and related steroids with rat thymus cells *in vitro*. *J. Biol. Chem.* **243**, 5556–5565.

49) Munck, A. and Foley, R. (1979) Activation of steroid hormone-receptor complexes in intact target cells in physiological conditions. *Nature* **278**, 752–754.

50) Murakami, N. and Moudgil, V.K. (1981) Extraction of DNA-cellulose-bound glucocorticoid-receptor complexes with sodium tungstate. *Biochim. Biophys. Acta* **676**, 386–394.
51) Murakami, N. and Moudgil, V.K. (1981) Inactivation of rat liver glucocorticoid receptor by molybdate. Effects on both non-activated and activated receptor complexes. *Biochem. J.* **198**, 447–455.
52) Murakami, N., Quattrociocchi, T.M., Healy, S.P., and Moudgil, V.K. (1982) Effects of sodium tungstate on the nuclear uptake of glucocorticoid-receptor complex from rat liver. *Arch. Biochem. Biophys.* **214**, 326–334.
53) Nielsen, C.J., Sando, J.J., and Pratt, W.B. (1977) Evidence that dephosphorylation inactivates glucocorticoid receptors. *Proc. Natl. Acad. Sci. U.S.A.* **74**, 1398–1402.
54) Nielsen, C.J., Sando, J.J., Vogel, W.M., and Pratt, W.B. (1977) Glucocorticoid receptor inactivation under cell-free conditions. *J. Biol. Chem.* **252**, 7568–7578.
55) Nishigori, H., Moudgil, V.K., and Toft, D.O. (1978) Inactivation of avian progesterone receptor binding to ATP-Sepharose by pyridoxal 5′-phosphate. *Biochem. Biophys. Res. Commun.* **80**, 112–118.
56) Okamoto, K., Isohashi, F., Horiuchi, M., and Sakamoto, Y. (1982) Three forms of macromolecular translocation inhibitor on the nuclear binding of "activated" receptor-glucocorticoid complex and their interaction with ATP. *Biochem. Biophys. Res. Commun.* **108**, 1655–1660.
57) Okamoto, K., Isohashi, F., Horiuchi, M., and Sakamoto, Y. (1984) An ATP-stimulated factor that enhances the nuclear binding of "activated" receptor-glucocorticoid complex. *Biochem. Biophys. Res. Commun.* **121**, 940–945.
58) Okamoto, K., Isohashi, F., Horiuchi, M., Yamamoto, A., Nakanishi, Y., and Sakamoto, Y. (1984) Purification and characterization of the ATP-stimulated translocation enhancing factor that is lacking in AH 130 tumor cells. Proc. 43rd Annual Meeting of the Japanese Cancer Association, Fukuoka, p. 331.
59) Okamoto, K., Isohashi, F., Tsukanaka, K., Horiuchi, M., and Sakamoto, Y. (1983) Alteration in the inducibility by dexamethasone of the activity of tyrosine aminotransferase and tryptophan oxygenase in regenerating rat liver after partial hepatectomy. *Endocrinology* **112**, 336–340.
60) Sakamoto, Y. and Isohashi, F. (1977) Specific receptors for glucocorticoids in tumor-bearing rats. In "Information on Receptors in Higher Animals," an Extra Edition of "Protein, Nucleic Acid and Enzyme," eds. Suda, M. and Nakagawa, H., pp. 181–189, Kyoritsu Shuppan, Tokyo (in Japanese).
61) Sakamoto, Y., Nakata, Y., and Isohashi, F. (1977) Transaminase activity and glucocorticoid in tumor-bearing rats. *Vitamins* **51**, 477–486.
62) Sando, J.J., Hammond, N.D., Stratford, C.A., and Pratt, W.B. (1979) Activa-

tion of thymocyte glucocorticoid receptors to the steroid binding form. *J. Biol. Chem.* **254**, 4779–4789.

63) Sando, J.J., La Forest, A.C., and Pratt, W.B. (1979) ATP-dependent activation of L cell glucocorticoid receptors to the steroid binding form. *J. Biol. Chem.* **254**, 4772–4778.

64) Scheidereit, C., Geisse, S., Westphal, H.M., and Beato, M. (1983) The glucocorticoid receptor binds to defined nucleotide sequences near the promoter of mouse mammary tumour virus. *Nature* **304**, 749–752.

65) Scheidereit, C., Krauter, P., Westphal, H.M., and Beato, M. (1984) Interaction of glucocorticoid hormone receptors with defined DNA sequences of inducible genes. *Prog. Cancer Res. Ther.* **31**, 89–96.

66) Schmidt, T.J., Bollum, F.J., and Litwack, G. (1982) Correlations between the activities of DNA polymerase α and the glucocorticoid receptor. *Proc. Natl. Acad. Sci. U.S.A.* **79**, 4555–4559.

67) Schmidt, T.J. and Litwack, G. (1982) Activation of the glucocorticoid-receptor complex. *Physiol. Rev.* **62**, 1132–1192.

68) Schmidt, T.J., Sekula, B.C., and Litwack, G. (1981) The effects of 1,10-phenanthroline on the binding of activated rat hepatic glucocorticoid-receptor complexes to deoxyribonucleic acid-cellulose. *Endocrinology* **109**, 803–812.

69) Schrader, W.T., Birnbaumer, M.E., Hughes, M.R., Weigel, N.L., Grody, W.W., and O'Malley, B.W. (1981) Studies on the structure and function of the chicken progesterone receptor. *Recent Prog. Hormone Res.* **37**, 583–633.

70) Sekula, B.C., Schmidt, T.J., and Litwack, G. (1981) Redefinition of modulator as an inhibitor of glucocorticoid receptor activation. *J. Steroid Biochem.* **14**, 161–166.

71) Sekula, B.C., Schmidt, T.J., Oxenham, E.A., DiSorbo, D.M., and Litwack, G. (1982) Dual effects of pyridoxal 5′-phosphate on glucocorticoid-receptor complexes. *Biochemistry* **21**, 2915–2922.

72) Simons, S.S. Jr. (1977) Evidence for nuclear and DNA binding forms of the receptor-glucocorticoid complex from hepatoma tissue culture cells. *Biochim. Biophys. Acta* **496**, 339–348.

73) Simons, S.S., Jr. (1979) Factors influencing association of glucocorticoid receptor-steroid complexes with nuclei, chromatin, and DNA: Interpretation of binding data. In "Glucocorticoid Hormone Action," eds. Baxter, J.D. and Rousseau, G.G., pp. 161–187, Springer-Verlag, Berlin, Heidelberg, New York.

74) Simons, S.S., Jr., Martinez, H.M., Garcea, R.L., Baxter, J.D., and Tomkins, G.M. (1976) Interactions of glucocorticoid receptor-steroid complexes with acceptor sites. *J. Biol. Chem.* **251**, 334–343.

75) Taira, M. and Terayama, H. (1978) Comparison of corticoid receptor and other

cytoplasmic factors among liver and hepatoma cell lines with different sensitivity to corticoid inhibition of cell growth. *Biochim. Biophys. Acta* **541**, 45–58.
76) Toft, D., Lohmar, P., Miller, J., and Moudgil, V. (1976) The properties and functional significance of ATP binding to progesterone receptors. *J. Steroid Biochem.* **7**, 1053–1059,
77) Yamamoto, A., Isohashi, F., Horiuchi, M., Okamoto, K., and Sakamoto, Y. (1984) Alteration in the level of ATP-stimulated translocation enhancer of the nuclear binding of activated-receptor glucocorticoid complex in regenerating livers of adrenalectomized rats after partial hepatectomy. *Seikagaku* **56**, 835 (in Japanese).

4

Glucocorticoid Hormone: Mechanisms of Action, ed. Sakamoto, Y. and Isohashi, F., pp. 97–116, Japan Sci. Soc. Press, Tokyo/Springer-Verlag, Berlin (1986)

BINDING SITES FOR THE GLUCOCORTICOID RECEPTOR ON HORMONALLY REGULATED GENES

M. BEATO, D. VON DER AHE, A.C.B. CATO, S. JANICH, P. KRAUTER, C. SCHEIDEREIT, G. SUSKE, M. WENZ, H.M. WESTPHAL, and T. WILLMANN

Institut für Physiologische Chemie, der Philipps-Universität, D-3550 Marburg, F.R.G.

Steroid hormones modulate the expression of different genetic programs in various target cells. The cellular specificity of the hormonal response depends upon the previous differentiation history of the cells, but in all cases the hormonal regulation of gene expression is mediated by specific proteins called receptors. The receptor for a particular steroid hormone appears to be the same in different cells of an organism and the variability in cellular response depends on the set of genes that is available for regulation in each type of cell. As in prokaryotic systems it seems reasonable to assume that the molecular mechanism of regulation will be similar for all cells and should involve an interaction of regulatory proteins with regulatory information encoded in the DNA.

In this paper we present evidence to the effect that the glucocorticoid receptor is indeed a regulatory protein that recognizes specific nucleotide sequences located at variable distances from glucocorticoid responsive promoters. In some cases we can show that the nucleotide sequences involved in this interaction are also required for hormone regulation and, thus, behave as glucocorticoid regulatory elements (GRE).

I. PURIFICATION AND CHARACTERIZATION OF THE RECEPTOR

A study of the interaction between the hormone receptor and regulated genes presupposes the purification and characterization of the receptor protein. We have used the rat liver as starting material for the purification, as the glucocorticoid receptor has been extensively characterized in this tissue (*7*, *13*, *39*, *77*). In fact, a correlation between receptor saturation and a physiological response was demonstrated first for the induction of liver enzymes by glucocorticoids in the rat (*8*).

The purification strategy is based on differential binding of the receptor to a polyanion (*17*). This procedure yields the activated form of the receptor, that can be used for DNA binding studies. If the starting material is frozen liver from adrenalectomized rats most of the receptor is recovered as a 40,000

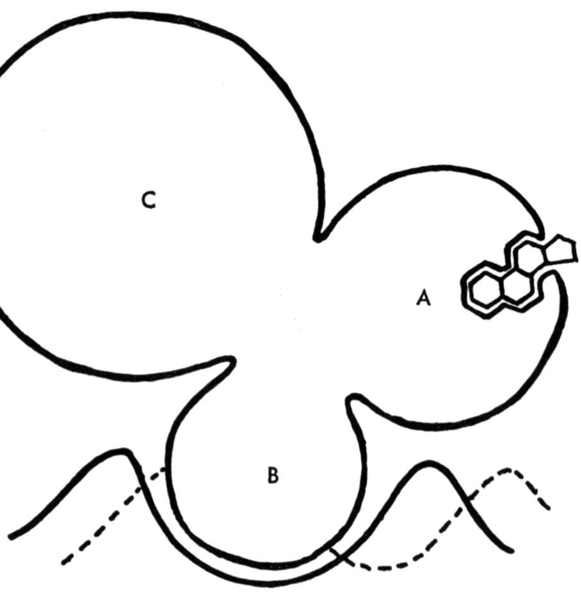

Fig. 1. Domain model of the 94,000 Mr form of the rat liver glucocorticoid receptor.

Mr form (*72*). If instead fresh liver is used, a larger 94,000 Mr form of receptor is obtained (*75*). This larger form, that probably represents the native polypeptide unit of the glucocorticoid receptor, can be cleaved by chymotrypsin into a 40,000 Mr form that still retains the steroid and DNA binding activities (*14*). The chymotryptic fragment appears to be equivalent to the 40,000 Mr form isolated from frozen liver (*72*). Identification of the various polypeptides as the steroid binding units has been achieved by photoaffinity labelling with the radioactive ligand (*73*).

The results of extensive proteolytic analysis of the native form of the receptor together with photoaffinity labelling and DNA binding studies (*78*) has led to the formulation of a domain model for the glucocorticoid receptor (Fig. 1). According to this model the basic receptor unit is composed of at least 3 domains: a domain A, that binds the steroid and has a molecular weight of Mr 25,000–27,000; a domain B that binds to DNA and has a MW of Mr 14,000; and a third domain (s) C, Mr 54,000 with more complex function. There is circumstantial evidence that the domain C may modulate the affinity of the B domain for the DNA by virtue of an interaction with other polypeptides (*18*, *53*).

There is also evidence that the glucocorticoid receptor in a phosphoprotein (*29*) and that phosphorylation may play a role in the metabolic cycle of the receptor (*5*). Phosphorylation appears to be required for steroid binding, but little is known about the influence of phosphorylation on DNA binding.

II. ANTIBODIES TO THE GLUCOCORTICOID RECEPTOR

Several groups have reported the isolation of polyclonal (*10*, *21*, *49*) and monoclonal (*24*, *27*, *50*, *74*) antibodies to the glucocorticoid receptor. In all cases, the antibodies were directed towards the C domain of the 94,000 Mr form of the receptor. In fact, antibodies have been used to detect this region of the protein for which no functional ligand is yet available. In Western blots with monoclonal receptor antibodies a main band at 94,000 Mr is found that comigrates with the steroid binding form of the receptor (*24*, *75*). For the first time this procedure offers the possibility to detect the receptor protein by means other than steroid binding. In fact, it could be shown that several mouse lymphoma cell lines with variant receptor phenotypes r^-, nt^i, and nt^- expressed a receptor polypeptide of 94,000 Mr that does not bind steroid (*75*).

The monoclonal antibodies are also useful to study the interaction of receptor with DNA in crude cytosol preparations (see below). The binding of the antibody to the C domain of the receptor apparently does not interfere with the interaction of the B domain with DNA (*65*).

Receptor antibodies have also been instrumental in the isolation and preliminary characterization of receptor cDNA and receptor genes (*44*). Ultimately these studies will lead to the elucidation of the complex receptor structure, and of the relationship between receptors for different hormones.

III. DNA BINDING ASSAYS

Until recently, DNA binding studies were performed with crude preparations of receptor and were based on rather indirect procedures (*35*, *47*, *55*). With the availability of purified receptor preparations and specific antibodies, more direct methods of analysis are possible.

The simplest test is based on the preferential retention of proteins on nitrocellulose filters (*60*). A set of radioactive DNA restriction fragments can be incubated with receptor preparations and filtered through nitrocellulose discs. Only those DNA fragments that bind to the receptor protein stick to the filters. The filters can then be washed, and the DNA eluted by detergents or high salt and analyzed by autoradiography after separation of the various fragments through gel electrophoresis. Using different restriction enzymes, this assay can serve to delimit the receptor binding site to less than a few hundred base pairs (*25*, *54*, *65*). In fact, if specific antibodies that do not interfere with DNA binding are available, a similar assay can be used with crude preparations of receptor. Instead of filtering through nitrocellulose discs, the DNA fragments bound to the receptor are isolated by immunological techniques (protein A or a second antibody) and analyzed by gel electrophoresis as in the filter assay (*65*).

A direct identification of the nucleotide sequences covered by the receptor can be accomplished in protection experiments against either exonuclease or endonucleases. Protection against DNase I, so-called footprint analysis, is a classical procedure that is based on the relative randomness of DNase I endonucleolytic degradation. Under appropriate conditions a sequence gel analysis of an end-labelled fragment after DNase I digestion will yield a continuous ladder of bands. If a protein is bound to a defined region of the DNA fragment the DNase I is not able to cut at these positions and no bands will be detected in this region of the sequencing gel (*23*). If a sequence reaction is run in parallel, this method allows a precise determination of the sequence covered by the receptor in a particular fragment (*54*, *65*). The limitation of this method is that rather pure preparations of receptor are needed, and the sensitivity is low; that means at least 60–70 % of the DNA binding sites have to be occupied in order to detect a clear footprint. These limitations can be circumvented by using specific antibodies to immunoisolate the protein-receptor complexes prior to DNase I digestion.

Exonuclease III protection experiments are less informative but more sensitive. Exonuclease III degrades double stranded DNA in a processive manner starting from each 3′ end (*68*). If a protein is bound to a defined region of the double strand, the enzyme will stop at the border of the region covered by the binding protein. If one 5′ end of the DNA is labelled, analysis of the digestion products in sequencing gels will show a radioactive band corresponding to the limit of the protected region (*68*). Contrary to the DNase I experiments that yield a negative signal (disappearance of bands), the exonuclease III protection experiments produce positive signals and are therefore more sensitive. The experiment can be performed with both strands, thus providing the limits of the footprint (*1*).

Information on the specific nucleotides involved in receptor binding and on the sites of contact between receptor and DNA can be obtained by chemical modification of individual nucleotides. The phosphate groups in direct contact with a DNA binding protein can be identified by protection against modification by ethylnitrose urea (*69*). Protection against methylation by dimethylsulfate allows identification of guanine and adenine residues that are in direct contact with a DNA binding protein (*48*). Conversely, the DNA bases can be specifically modified, and the influence of this chemical change on protein binding can be tested. When methylation by dimethylsulfate is used, the procedure is called methylation interference (*69*). All these procedures are based on the use of DNA fragments labelled at one end, followed by analysis of the modified and cleaved DNA fragments in sequencing gels.

IV. DNA BINDING SITES FOR THE GLUCOCORTICOID RECEPTOR

Several hormonally regulated genes have been explored for the presence of DNA sequences that are recognized by the activated glucocorticoid receptor of rat liver. In the following we will summarize the results obtained in our laboratory and by other groups with the following genes: mouse mammary tumor virus, human metallothionein II_A, chicken lysozyme, rabbit uteroglobin, and human growth hormone.

1. Mouse Mammary Tumor Virus (MMTV)

The induction of MMTV-RNA by glucocorticoids has been used for many years as a model for gene regulation by steroid hormones (*61*). In fact, this was among the first cases in which gene transfer experiments demonstrated that the information required for hormonal regulation is located in a fragment of cloned DNA, namely, within the long terminal repeat (LTR) region (*11*, *30*, *31*, *40*). Several groups then showed a receptor dependent preferential retention of specific restriction fragments of the MMTV provirus in filter binding experiments (*25*, *26*, *52*, *54*, *55*). Preferential binding of the receptor was consistently detected in the LTR-region and occasionally in the *env*-gene region and in the mouse flanking DNA (*25*).

A fine deletion analysis of the LTR-region in gene transfer experiments delimited the DNA sequences responsible for glucocorticoid inducibility between positions −50 and −200 upstream of the transcription initiation site (*12*, *32*, *41*). In fact, these regions appear to be composed of several elements with different biological efficiency, with the main element located between −140 and −190. In agreement with these biological data, DNase I footprint experiments demonstrated the existence of four binding sites for the glucocorticoid receptor in the LTR region between −71 and −192, with the highest affinity site located between −163 and −192 (*65*). Common to all four binding sites is the hexanucleotide 5′-TGTTCT-3′, but additional sequence homologies are found between the individual sites (*65*). Using different strains of MMTV, similar results have been obtained by other groups (*51*, *56*), who also found additional receptor sites further upstream in the LTR.

The sites of contact between the receptor and the DNA within the binding sites were analyzed by methylation interference and methylation protection experiments (*69*). In addition to the two G-residues in the hexanucleotide
5′-TGTTCT-3′
3′-ACTAGA-5′ that were protected in all four binding sites, two other G-residues some 9–10 base pairs (bp) upstream are also protected in the strong binding site at −180 and −184 (*64*). These four G-residues appear to be essential for receptor recognition (see below).

The combination of gene transfer experiments and receptor binding studies with MMTV served to identify for the first time a regulatory element for glucocorticoids, abbreviated GRE: a set of DNA sequences that is involved in receptor binding and mediates hormonal regulation of gene expression.

2. Human Metallothionein II_A Gene (hMT II_A)

Another family of glucocorticoid inducible genes that have been extensively studied in gene transfer experiments are the metallothionein genes. Whereas the mouse or human metallothionein I genes are not inducible by glucocorticoids after transfer of cloned genes in host cells (*42*, *59*), successful experiments have been reported with the cloned hMT II_A (*36*, *38*). In a deletion analysis, the sequences responsible for glucocorticoid inducibility were localized between positions −236 and −268 (*37*). Both filter binding experiments and DNase I footprint analysis demonstrated the existence of a main binding site for the glucocorticoid receptor located between −242 and −267 (*37*). Thus, these experiments demonstrated the presence of a glucocorticoid regulatory element composed of a single receptor binding site in a non-viral inducible gene. In fact, another weak binding site was found between −318 and −332, but this site could be deleted without reducing hormone inducibility (*37*). Common to both sites was the hexanucleotide 5′-TGTCCT-3′, similar to the one found in the MMTV-LTR, and methylation protection experiments showed similar receptor contact sites in the hMT II_A promoter and in the MMTV-LTR region (*37*, *66*). Thus, it appears that the rat liver glucocorticoid receptor interacts with regulatory elements from mouse and human in essentially the same way.

3. Chicken Lysozyme Gene

The genes for egg-white proteins are expressed in the chick oviduct under the control of various steroid hormones including glucocorticoids (*28*, *45*). Gene transfer experiments with the lysozyme gene failed to show expression of this gene in a variety of cells with the exception of primary oviduct cell cultures (*57*). When the lysozyme promoter is linked to the SV40 T antigen and microinjected in oviduct cells, both glucocorticoid and progesterone regulation of T antigen expression is observed (*57*). Deletion analysis showed that sequences between −164 and −208 are required for hormonal regulation (*58*). Filter binding experiments and exonuclease III footprint analysis identified two glucocorticoid receptor binding sites, one located between −174 and −200, and the other between −50 and −84 (*1*, *58*). The upstream binding site exhibits a low affinity for the glucocorticoid receptor and weak homology to the strong binding sites found in MMTV-LTR and hMT II_A. Since, however, deletion of this site leads to loss of hormonal regulation, these sequences must be part of a glucocorticoid regulatory element. The binding site between −50 and −84 exhibits a higher affinity for the glucocorticoid receptor and contains the hexanucleotide 5′-TGTTCT-3′ in the lower strand. It is not known whether alteration of this site alone could also result in loss of hormonal regulation. If, however, this element plays a role in hormonal regulation, this would mean that the glucocorticoid regulatory elements can act in both orientations.

4. *Rabbit Uteroglobin Gene (RUG)*

The rabbit uteroglobin gene is expressed in the lung under the control of glucocorticoids (*22*) and in endometrium under the control of ovarian hormones (*6*). Since the gene and its flanking DNA have been cloned and sequenced (*43*, *70*), we have searched for glucocorticoid binding sites around the uteroglobin promoter. To our surprise we were not able to detect any receptor binding site within a few hundred base pairs from the initiation of transcription. In fact, the only sites with high affinity for the glucocorticoid receptor were located over 2,600 bp upstream of the "cap"-site (*15*). In filter binding experiments, DNase I footprint analysis and methylation protection experiments, we mapped three contiguous binding sites with the hexanucleotide $5'\text{-}{}^{\text{A}}_{\text{T}}\text{GT}^{\text{T}}_{\text{C}}\text{CT-}3'$ between −2,619 and −2,708 (*15*). The contact sites between the receptor and the DNA double helix within the binding region are very similar to those found in MMTV-LTR and hMT II_A (Fig. 2). It has not been proven that these sites are actually responsible for the glucocorticoid induction of uteroglobin gene expression, but the lack of other receptor binding sites closer to the uteroglobin promoter makes this very probable. In that case, the glucocorticoid regulatory elements will be active even several kilobases (kb) away from the regulated promoter.

5. *Human Growth Hormone Gene (hGHI)*

The expression of the hGHI gene is induced by glucocorticoid and thyroid hormones in pituitary cells (*3*). There are reports in the literature indicating

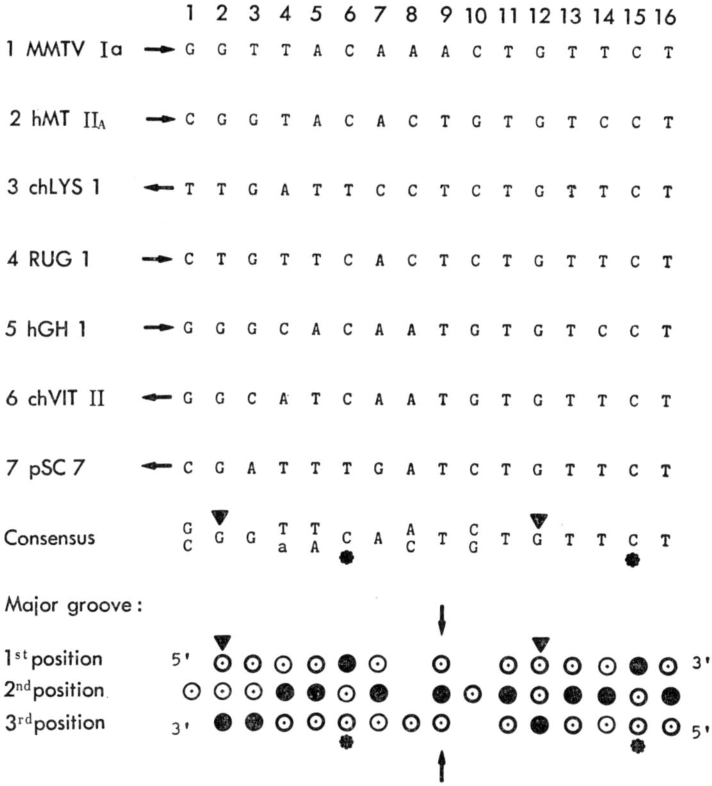

Fig. 2. Nucleotide sequence of the binding sites for the glucocorticoid receptor in the following genes: 1, mouse mammary tumor virus (MMTV Ia); 2, human metallothionein II_A (hMT II_A); 3, chicken lysozyme (chLYS 1); 4, rabbit uteroglobin (RUG 1); 5, human growth hormone (hGH I); 6, chicken vitellogenin (chVIT II); and 7, *Drosophila* salivary gland protein (pSC 7). The arrows indicate whether the sequence is in the upper strand (→) or in the lower strand (←). The G-residues protected by the receptor against methylation in the upper (▼) and lower (✻) strands are indicated. In the lower part of the figure, the pattern of conserved hydrogen bond donor (●) and acceptor (⊙) positions in the major groove is shown. An inverted symmetry centered around position 9 is evident.

that the 5′-flanking sequences of the hGHI can confer glucocorticoid inducibility over the linked thymidine kinase gene (*62*). In binding experiments, however, we could not find a tight binding of the glucocorticoid receptor to the 5′-upstream region of the gene (unpublished). Instead, a binding site was found between +86 and +114 within the first intron of hGHI (*19* and unpublished results). This site, which contains the hexanucleotide 5′-TGTCCT-3′, can replace the glucocorticoid regulatory element of $hMTII_A$ gene in gene transfer experiments (unpublished). The definitive proof of an involvement of the intron sequences in the hormonal regulation of hGHI expression awaits further experiments. If, however, the available preliminary data are confirmed, they would suggest that the glucocorticoid regulatory elements can naturally operate downstream from the regulated promoter. In artificial construction and gene transfer experiments similar results have been obtained with MMTV-LTR (*16*).

V. RELATIONSHIP BETWEEN "GRE" AND BINDING SITES FOR OTHER STEROID HORMONE RECEPTORS

Many of the hormonally regulated genes can be induced by different steroid hormones. For instance, the expression of the lysozyme gene in chick oviduct can be induced by estrogens, progesterone, glucocorticoids, and androgens (*45*). We were interested to know whether individual regulatory elements exist for each steroid hormone or whether all hormones act through similar or identical regulatory elements. An indication that the regulatory elements for glucocorticoids and progesterone may overlap in the chicken lysozyme promoter came from the observation that deletion of only 44 bp between −164 and −208 leads to loss of inducibility by both hormones (*58*).

We investigated the binding of the rabbit progesterone receptor to the chicken lysozyme promoter region and found two binding sites (*1*). One site coincides exactly with the proximal binding site for the glucocorticoid receptor between −50 and −84, whereas the other binding site extends from −160 to −200, overlapping but not exactly coinciding with the distal binding site for the glucocorticoid receptor (*1*). In addition, the relative affinity of the progesterone and the glucocorticoid receptors for these two sites were reversed, the distal binding site having higher affinity for the progesterone receptor while the proximal binding site showed higher affinity for the glucocorticoid receptor. Thus, although the regulatory elements for both hormones overlap, the molecular mechanism of the interaction between the two receptors and DNA may differ.

The rabbit progesterone receptor also binds to the four sites in MMTV-LTR that are recognized by the rat glucocorticoid receptor (*1*). Although the limits of the exonuclease III footprints are identical for both receptors, the relative affinities for individual sites differ. As for the functional significance of these findings in relation to hormonal regulation of MMTV expression, a search of the literature did not show any effect of progesterone. Preliminary gene transfer experiments, however, indicate that the receptor binding sites in the LTR-region can confer progesterone inducibility upon a thymidine kinase promoter located downstream (unpublished).

Although there is a report on preferential binding of the progesterone receptor to the promoter region of the uteroglobin gene (*2*), we have not been able to confirm this finding. Instead, in filter binding experiments we have found that the progesterone receptor binds to the same restriction fragment of the upstream uteroglobin region that is bound by the glucocorticoid receptor (unpublished).

Thus, it is possible that both glucocorticoids and progesterone regulate uteroglobin gene expression by means of an interaction of their receptors with sequences located around 2.6 kb upstream of the uteroglobin promoter.

The chicken vitellogenin II gene has been reported to contain a binding site for the estrogen receptor located about 600 bp upstream of the transcription initiation site (*33*). Overlapping this region we found a binding site for the rat glucocorticoid receptor with the hexanucleotide 5′-TGTTCT-3′ in the lower strand (Fig. 2; unpublished). Nothing is known about the physiological significance of this finding, but it clearly shows that the receptors for estrogen and glucocorticoids interact with overlapping DNA sequences.

Recently an ecdysone inducible gene from *Drosophila* has been cloned and partly sequenced (*46*). In the 5′-flanking sequence of this gene we found a binding site for the rat glucocorticoid receptor containing the hexanucleotide 5′-TGTTCT-3′ at position −304 in the lower strand (unpublished). It is not known whether the ecdysone receptor interacts with adjacent sequences, but our finding is compatible with the idea that the binding sites for different steroid hormone receptors may share structural features.

VI. ROLE OF THE STEROID LIGAND

One important question concerning the interaction of the receptor with DNA is whether the steroid ligand is required for specific DNA binding. There are early literature reports indicating that hormone binding is a prerequisite for receptor activation (*9*, *34*), but at that time the receptor was measured by virtue of its binding to radioactive steroid. With the help of monoclonal antibodies that do not interfere with DNA binding (*65*, *74*), we can now question the role of the ligand in a more precise way. Using crude cytosol and a set of radioactive restriction fragments from MMTV, we observed a preferential binding of receptor to the LTR-region independent of the addition of saturating amounts of triamcinolone acetonide or corticosterone (unpublished). In fact, even in the presence of an antiglucocorticoid, RU486, the cytosol receptor bound specifically to the LTR fragments. DNase I protection experiments demonstrate that similar footprints were obtained under all these conditions, and these footprints were comparable to those produced by purified receptor (unpublished). Thus, it seems that in the absence of the ligand the receptor can recognize similar DNA sequences as those bound by the purified glucocorticoid-receptor complex.

In high resolution Western blot analysis of crude liver cytosol we can detect two polypeptide bands in the 90–94 kilodalton (Kd) region that react with the monoclonal antibody. It seems that only the largest of these two bands can bind steroid, whereas the faster band is not retained by a steroid-matrix. Therefore, it is conceivable that this faster population of immunoreactive polypeptides is responsible for the steroid independent binding to specific DNA sequences. If this is the case, these polypeptides probably represent a population of inactive receptor molecules unable to bind the steroid hormone. The very interesting question then arises of whether these steroid-free receptor molecules could be already bound to their target sites on DNA before the hormone comes into play. The function of the hormone would then be to activate an event distal to the recognition of specific DNA sequences.

VII. MODEL OF DNA-RECEPTOR INTERACTION

A comparison of the strong binding sites for the glucocorticoid receptor reported in this paper allows derivation of a consensus sequence for the GRE with an axis of symmetry at position 9 (Fig. 2). Certain positions such as 11, 12, 13, 15, and 16 are 100 % preserved, other positions such as 5, 8, 10, and 14 accept only two bases: in 5, A or T, in 8, A or C, in 10, C or G, and in 14, T or C. Position 9 is T in 6 cases, and other positions such as 2, 6, and 7 are preserved in 5 out of 7 sequences. The sites of contact between receptor and DNA, as determined by methylation protection experiments (*64*), are the N-7 position of the following G-residues: 2 and 12 in the upper strand and 6 and 15 in the lower strand. Since these contact sites are separated by 10 and 9 bp respectively, the data are compatible with a model in which two receptor molecules contact two consecutive turns of the double helix, approaching the DNA through the major groove (*64*).

The recognition of specific base pairs should include hydrogen bonding between amino acid residues and bases on the DNA. Each base pair has 3 hydrogen bonding positions. In respect to the pattern of hydrogen bond donor and acceptor sites, an A-T base is equivalent to a T-A base pair; a C-G and a G-C pair base share the central acceptor site, whereas an A-T and a C-G base pair share the 3rd hydrogen acceptor site (*63*, *67*). Taking this into account, a conserved pattern of potential hydrogen bonding sites can be derived from the binding sites listed in Fig. 2, that allows one to envisage how the receptor could recognize the DNA double helix within the regulatory elements (Fig. 2). This is, of course, an oversimplification, since the exact positions of the hydrogen bonding sites in different base pairs is not identical (*20*) and steric hindrance by the methyl group at thymidine has not been considered. Nevertheless, this type of analysis could help to understand the mechanisms by which regulatory information encoded in the DNA is read by DNA binding regulatory proteins.

SUMMARY

The following conclusions can be drawn from the results summarized in this paper.

1. Glucocorticoid regulated genes contain binding sites for the glucocorticoid receptor located at variable distances from the transcription initiation site: from more than 2,600 bp upstream to around 100 bp downstream. It also seems probable that the GRE can act in both orientations. These properties are reminiscent of those reported for so-called "enhancer" elements that have been found in a variety of viral and cellular genes (*4*, *71*). Thus, it is conceivable that the GRE behave as hormone dependent enhancers. Since it is believed that enhancer elements act as binding sites for specific DNA binding proteins, the glucocorticoid receptor would be the first of these proteins that has been identified.

2. The sequence of these GRE appears to have been highly conserved between chicken, rodent, and human genes, as they are all recognized by the rat glucocorticoid receptor.

3. The receptors for different steroid hormones interact with similar, or at least overlapping, DNA sequences. This may explain why many genes are known to be regulated by different steroid hormones, and why deletion of certain sequences leads to concomitant loss of inducibility by different hormones (*58*). It is therefore probable that the DNA binding domains of different steroid hormone receptors are similar.

4. Binding of the steroid ligand to the receptor is not an absolute requirement for a specific interaction of the receptor with DNA. Apparently there is a population of receptor molecules that does not bind the steroid but still is able to interact with DNA. Thus, it is conceivable that receptor molecules are already associated with their target sequences on DNA before the hormone reaches the cell.

5. The question is still open why in different cells different genes are inducible by the same hormone presumably through the same receptor. Other chromosomal proteins may be involved in determining tissue specific accessibility of different genes, or additional tissue specific proteins may be needed for the receptor binding to result in regulated transcription. This question, that ultimately relates to the mechanism of cell differentiation, will attract our attention during the next decade.

Acknowledgments
We thank N. Hynes and B. Groner (Bern, Switzerland), M. Karin (Los Angeles, U.S.A.), R. Renkawitz and G. Schütz (Heidelberg, FRG) for providing some of the recombinant plasmids, and Mrs. H. Ballach for the preparation of the manuscript. The experimental work was supported by grants from the Deutsche Forschungsgemeinschaft, the Kempkes-Stiftung and the Fonds der Chemischen Industrie.

REFERENCES

1) Ahe, D. von der, Janich, S., Scheidereit, C., Renkawitz, R., Schütz, G., and Beato, M. (1985) The glucocorticoid and the progesterone receptors bind to the same sites in two hormonally regulated promoters. *Nature* **313**, 706–709.
2) Bailley, A., Atger, M., Atger, P., Cerbon, M.-A., Alizon, M., Hai, M.T. vu,

Logeat, F., and Migrom, E. (1983) The rabbit uteroglobin gene. Structure and interaction with the progesterone receptor. *J. Biol. Chem.* **258**, 10384–10389.

3) Bancroft, F.C., Levine, L., and Tashjian, A.H., Jr. (1969) Control of growth hormone production by clonal strain of rat pituitary cells. Stimulation by hydrocortisone. *J. Cell. Biol.* **43**, 432–441.
4) Banerji, J., Rusconi, S., and Schaffner, W. (1981) Expression of a β-globin gene is enhanced by remote SV40 DNA sequences. *Cell* **27**, 299–308.
5) Barnett, C.A., Schmidt, T.J., and Litwack, G. (1980) Effect of calf intestinal alkaline phosphatase, phosphatase inhibitors and phosphorylated compounds on the rate of activation of glucocorticoid-receptor complexes. *Biochemistry* **19**, 5446–5455.
6) Beato, M., Arnemann, J., Menne, C., Müller, H., Suske, G., and Wenz, M. (1983) Regulation of the expression of the uteroglobin gene by ovarian hormones. In "Regulation of Gene Expression by Hormones," ed. McKerns, K.W., pp. 151–175, Plenum Press, New York.
7) Beato, M. and Feigelson, P. (1972) Glucocorticoid-binding proteins of rat liver cytosol. I. Separation and identification of the binding proteins. *J. Biol. Chem.* **247**, 7890–7896.
8) Beato, M., Kalimi, M., and Feigelson, P. (1972) Correlation between glucocorticoid binding to specific liver cytosol receptors and enzyme induction *in vivo*. *Biochem. Biophys. Res. Commun.*, **47**, 1464–1472.
9) Beato, M., Kalimi, M., Konstam, M., and Feigelson, P. (1973) Interaction of glucocorticoids with rat liver nuclei: II. Studies on the nature of the cytosol transfer factor and the nuclear acceptor site. *Biochemistry* **12**, 3372–3379.
10) Bernard, P.A. and Joh, T.H. (1984) Characterization and immunocytochemical demonstration of glucocorticoid receptor using antisera specific to transformed receptor. *Arch. Biochem. Biophys.* **229**, 466–476.
11) Buetti, E. and Diggelmann, H. (1981) Cloned MMTV virus DNA is biologically active in transfected mouse cells and its expression is stimulated by glucocorticoid hormones. *Cell* **23**, 335–345.
12) Buetti, E. and Diggelmann, H. (1983) Glucocorticoid regulation of MMTV: identification of a short essential region. *EMBO J.* **2**, 1423–1429.
13) Cake, M.H. and Litwack, G. (1975) The glucocorticoid receptor. In "Biochemical Action of Hormones," Vol. III, ed. Litwack, G., pp. 317–390, Academic Press, New York.
14) Carlstedt-Duke, J., Okret, S., Wrange, Ö., and Gustafsson, J.-A. (1982) Immunochemical analysis of the glucocorticoid receptor: identification of a third domain separated from the steroid-binding and DNA-binding domains. *Proc. Natl. Acad. Sci. U.S.A.* **79**, 4260–4264.
15) Cato, A.C.B., Geisse, S., Wenz, M., Westphal, H.M., and Beato, M. (1984)

The nucleotide sequences recognized by the glucocorticoid receptor in the rabbit uteroglobin gene are located far upstream from the initiation of transcription. *EMBO J.* **3**, 2771–2778.
16) Chandler, V.L., Maler, B.A., and Yamamoto, K.R. (1983) DNA sequences bound specifically by glucocorticoid receptor *in vitro* render a heterologous promoter hormone responsive *in vivo*. *Cell* **33**, 489–499.
17) Climent, F., Bugany, H., and Beato, M. (1976) Partial purification of the activated glucocorticoid receptor of rat liver. *FEBS Lett.* **66**, 317–321.
18) Dellweg, H.-G., Hotz, A., Mugele, K., and Gehring, U. (1982) Active domains in wild-type and mutant glucocorticoid receptors. *EMBO J.* **1**, 285–289.
19) DeNoto, F.M., Moore, D.D., and Goodman, H.M. (1981) Human growth hormone DNA sequence and mRNA structure: possible alternative splicing. *Nucleic Acids Res.* **15**, 3719–3730.
20) Dickerson, R.E. and Drew, H.R. (1981) Structure of a B-DNA dodecamer. II. Influence of base sequence on helix structure. *J. Mol. Biol.* **149**, 761–786.
21) Eisen, H.J. (1980) An antiserum to the rat liver glucocorticoid receptor. *Proc. Natl. Acad. Sci. U.S.A.* **77**, 3893–3897.
22) Fernandez-Renau, D., Lambardero, M., and Nieto, A. (1984) Glucocorticoid-dependent uteroglobin synthesis and uteroglobin mRNA levels in rabbit lung explants cultured *in vitro*. *Eur. J. Biochem.* **144**, 523–527.
23) Galas, D.J. and Schmitz, A. (1978) DNase footprinting: a simple method for the detection of protein-DNA binding specificity. *Nucleic Acids Res.* **5**, 3157–3170.
24) Gametchu, B. and Harrison, R.W. (1984) Characterization of a monoclonal antibody to the rat liver glucocorticoid receptor. *Endocrinology* **114**, 274–279.
25) Geisse, S., Scheidereit, C., Westphal, H.M., Hynes, N.E., Groner, B., and Beato, M. (1982) Glucocorticoid receptors recognize DNA sequences in and around murine mammary tumour virus DNA. *EMBO J.* **1**, 1613–1619.
26) Govindan, M.V., Spiess, E., and Majors, J. (1982) Purified glucocorticoid receptor-hormone complex from rat liver cytosol binds specifically to cloned mouse mammary tumor virus long terminal repeats *in vitro*. *Proc. Natl. Acad. Sci. U.S.A.* **79**, 5157–5161.
27) Grandics, P., Gasser, D.L., and Litwack, G. (1982) Monoclonal antibodies to the glucocorticoid receptor. *Endocrinology* **111**, 1731–1733.
28) Hager, L.J., McKnight, G.S., and Palmiter, R.D. (1980) Glucocorticoid induction of egg white mRNAs in chick oviduct. *J. Biol. Chem.* **255**, 7796–7800.
29) Housley, P.R. and Pratt, W.B. (1983) Direct demonstration of glucocorticoid receptor phosphorylation by intact L-cells. *J. Biol. Chem.* **258**, 4630–4635.
30) Huang, A.L., Ostrowski, M.C., Berard, D., and Hager, G.L. (1981) Gluco-

corticoid regulation of the Ha-MuSVp21 gene conferred by sequences from mouse mammary tumor virus. *Cell* **27**, 245–255.

31) Hynes, N.E., Kennedy, N., Rahmsdorf, U., and Groner, B. (1981) Hormone responsive expression of an endogenous proviral gene of MMTV after molecular cloning and gene transfer into cultured cells. *Proc. Natl. Acad. Sci. U.S.A.* **78**, 2038–2042.
32) Hynes, N.E., van Ooyen, A.J.J., Kennedy, N., Herrlich, P., Ponta, H., and Groner, B. (1983) Subfragments of the large terminal repeat cause glucocorticoid responsive expression of mouse mammary tumor virus and of an adjacent gene. *Proc. Natl. Acad. Sci. U.S.A.* **80**, 3637–3641.
33) Jost, J.-P., Seldran, M., and Geiser, M. (1984) Preferential binding of estrogen-receptor complex to a region containing the estrogen-dependent hypomethylation site preceding the chicken vitellogenin II gene. *Proc. Natl. Acad. Sci. U.S.A.* **81**, 429–433.
34) Kalimi, M., Beato, M., and Feigelson, P. (1973) Interaction of glucocorticoid with rat liver nuclei. I. Role of the cytosol proteins. *Biochemistry* **12**, 3365–3361.
35) Kallos, J. and Hollander, V.P. (1978) Assessment of specificity of estrogen receptor-DNA interaction by a competitive assay. *Nature* **272**, 177–179.
36) Karin, M., Haslinger, A., Holtgreve, H., Cathala, G., Slater, E., and Baxter, J.D. (1984) Activation of a heterologous promoter in response to dexamethasone and cadmium by metallothionein gene 5′-flanking DNA. *Cell* **36**, 371–379.
37) Karin, M., Haslinger, A., Holtgreve, H., Richards, R.I., Krauter, P., Westphal, H.M., and Beato, M. (1984) Characterization of DNA sequences through which cadmium and glucocorticoid hormones induce human metallothionein II_A-gene. *Nature* **308**, 513–519.
38) Karin, M. and Richards, R.I. (1982) Human metallothionein genes—primary structure of the metallothionein II gene and a related processed gene. *Nature* **299**, 797–802.
39) Koblinski, M., Beato, M., Kalimi, M., and Feigelson, P. (1972) Glucocorticoid-binding proteins of rat liver cytosol. II. Physical characterization and properties of the binding proteins. *J. Biol. Chem.* **247**, 7897-7904.
40) Lee, F., Mulligan, R., Berg, O., and Ringold, G. (1981) Glucocorticoid regulates expression of dihydrofolate reductase cDNA in mouse mammary tumor virus chimeric plasmids. *Nature* **294**, 228–232.
41) Majors, J. and Varmus, H.E. (1983) A small region of mouse mammary tumor virus long terminal repeat confers glucocorticoid hormone regulation on a linked heterologous gene. *Proc. Natl. Acad. Sci. U.S.A.* **80**, 5866–5870.
42) Mayo, K.E., Warren, R., and Palmiter, R.D. (1982) The mouse metallothio-

nein-I gene is transcriptionally regulated by cadmium following transfection into human or mouse cells. *Cell* **29**, 99–108.

43) Menne, C., Suske, G., Arnemann, J., Wenz, M., Cato, A.C.B., and Beato, M. (1982) Isolation and structure of the gene for the progesterone inducible protein uteroglobin. *Proc. Natl. Acad. Sci. U.S.A.* **79**, 4853–4857.

44) Miesfeld, R., Okret, S., Wikstrom, A.C., Wrange, Ö., Gustafsson, J.-A., and Yamamoto, K.R. (1984) Initial characterization of the glucocorticoid receptor gene and transcripts in wild type and mutant cells using cloned cDNA. *Nature* **312**, 779–781.

45) Moen, R.C. and Palmiter, R.D. (1980) Changes in hormone responsiveness of chick oviduct during primary stimulation with estrogen. *Dev. Biol.* **78**, 450–463.

46) Moritz, T., Edström, J.E., and Pongs, O. (1984) Cloning of a gene localized and expressed at the ecdysteroid regulated puff 74EF in salivary glands of *Drosophila* larvae. *EMBO J.* **3**, 289–295.

47) Mulvihill, E.R., Lepennec, J.P., and Chambon, P. (1982) Chicken oviduct progesterone receptor: location of specific regions of high affinity binding in cloned DNA fragments of hormone responsive chicken genes. *Cell* **28**, 621–632.

48) Ogata, R.T. and Gilbert, W. (1978) An amino-terminal fragment of the *lac* repressor binds specifically to *lac* operator. *Proc. Natl. Acad. Sci. U.S.A.* **75**, 5851–5854.

49) Okret, S., Carlstedt-Duke, J., Wrange, Ö., Carlström, K., and Gustafsson, J.-A. (1981) Characterization of an antiserum against the glucocorticoid receptor. *Biochim. Biophys. Acta* **677**, 205–219.

50) Okret, S., Wikström, A.-Ch., Wrange, Ö., Anderson, B., and Gustafsson, J.-A. (1984) Monoclonal antibodies against the rat liver glucocorticoid receptor. *Proc. Natl. Acad. Sci. U.S.A.* **81**, 1609–1613.

51) Payvar, F., DeFranco, D., Firestone, G.L., Edgar, B., Wrange, Ö., Okret, S., Gustafsson, J.-A., and Yamamoto, K.R. (1983) Sequence-specific binding of glucocorticoid-receptor to MMTV DNA at sites within and upstream of the transcribed region. *Cell* **35**, 381–392.

52) Payvar, F., Firestone, G., Ross, S.R., Chandler, V.L., Wrange, Ö., Carlstedt-Duke, J., Gustafsson, J.-A., and Yamamoto, K.R. (1982) Multiple specific binding sites for purified glucocorticoid receptors on mammary tumor virus DNA. *J. Cell. Biochem.* **19**, 241–247.

53) Payvar, F. and Wrange, Ö. (1983) Relative selectivities and efficiencies of DNA binding by purified intact and protease-cleaved glucocorticoid receptor. In "Steroid Hormone Receptors: Structure and Function, Nobel Symp. 57," eds.

Eriksson, H., Gustafsson, J.-A., and Högberg, B., pp. 267–282, Elsevier/North Holland, Biomed. Press, Amsterdam.

54) Payvar, F., Wrange, Ö., Carlstedt-Duke, J., Okret, S., Gustafsson, J.-A., and Yamamoto, K.R. (1981) Purified glucocorticoid receptors bind selectively *in vitro* to a cloned DNA fragment whose transcription is regulated by glucocorticoid *in vivo*. *Proc. Natl. Acad. Sci. U.S.A.* **78**, 6628–6632.
55) Pfahl, M. (1982) Specific binding of the glucocorticoid-receptor complex to the mouse mammary tumor proviral promoter region. *Cell* **31**, 475–482.
56) Pfahl, M., McGinnis, D., Hendricks, M., Groner, B., and Hynes, N.E. (1983) Correlation of glucocorticoid receptor binding sites on MMTV proviral DNA with hormone inducible transcription. *Science* **222**, 1341–1343.
57) Renkawitz, R., Beug, H., Graf, T., Matthias, P., Grez, M., and Schütz, G. (1982) Expression of a chicken lysozyme recombinant gene is regulated by progesterone and dexamethasone after microinjection in oviduct cells. *Cell* **31**, 167–176.
58) Renkawitz, R., Schütz, G., von der Ahe, D., and Beato, M. (1984) Identification of hormone regulatory elements in the promoter region of the chicken lysozyme gene. *Cell* **37**, 503–510.
59) Richards, R.I., Hegury, A., and Karin, M. (1984) Structural and functional analysis of the human metallothionein-I_A gene: differential induction by metal ions and glucocorticoids. *Cell* **37**, 263–272.
60) Riggs, A.D., Suzuki, H., and Bourgeois, S. (1970) Lac repressor-operator interaction. I. Equilibrium studies. *J. Mol. Biol.* **48**, 67–83.
61) Ringold, G.M., Yamamoto, K.R., Tomkins, G.M., Bishop, J.M., and Varmus, H.E. (1975) Dexamethasone-mediated induction of mouse mammary tumor virus RNA: a system for studying glucocorticoid action. *Cell* **6**, 299–305.
62) Robins, D.M., Paek, I., Seeburg, P., and Axel, R. (1982) Regulated expression of human growth hormone genes in mouse cells. *Cell* **29**, 623–631.
63) Sadler, J.R., Waterman, M.S., and Smith, T.F. (1983) Regulatory pattern identification in nucleic acid sequences. *Nucleic Acids Res.* **11**, 2221–2231.
64) Scheidereit, C. and Beato, M. (1984) Contacts between receptor and DNA double helix within a glucocorticoid regulatory element of mouse mammary tumor virus. *Proc. Natl. Acad. Sci. U.S.A.* **81**, 3029–3034.
65) Scheidereit, C., Geisse, S., Westphal, H.M., and Beato, M. (1983) The glucocorticoid receptor binds to defined nucleotide sequences near the promoter of mouse mammary tumor virus. *Nature* **304**, 749–752.
66) Scheidereit, C., Krauter, O., Westphal, H.M., and Beato, M. (1984) Interaction of glucocorticoid hormone receptors with defined DNA sequences of inducible genes. In "Hormones and Cancer," Vol. 2, eds. Brescianie, F. *et al.*, pp. 89–96, Raven Press, New York.

67) Seeman, N.C., Rosenberg, J.M., and Rich, A. (1976) Sequence-specific recognition of double helical nucleic acids by proteins. *Proc. Natl. Acad. Sci. U.S.A.* **73**, 804–808.

68) Shalloway, D., Kleinberger, T., and Livingston, D.M. (1980) Mapping of SV40 DNA replication origin region binding sites for the SV40 T antigen by protection against exonuclease III digestion. *Cell* **20**, 411–422.

69) Siebenlist, U. and Gilbert, W. (1980) Contacts between *E. coli* RNA polymerase and an early promoter of phage T7. *Proc. Natl. Acad. Sci. U.S.A.* **77**, 122–126.

70) Suske, G., Wenz, M., Cato, A.C.B., and Beato, M. (1983) The uteroglobin gene region: hormonal regulation, repetitive elements and complete nucleotide sequence of the gene. *Nucleic Acids Res.* **11**, 2257–2271.

71) Weiher, H., König, M., and Gruss, P. (1983) Multiple point mutations affecting the SV40 enhancer. *Science* **219**, 626–631.

72) Westphal, H.M. and Beato, M. (1980) The activated glucocorticoid receptor of rat liver. Purification and physical characterization. *Eur. J. Biochem.* **106**, 395–403.

73) Westphal, H.M., Fleischmann, G., and Beato, M. (1981) Photoaffinity labelling of steroid binding proteins with unmodified ligands. *Eur. J. Biochem.* **119**, 101–106.

74) Westphal, H.M., Moldenhauer, G., and Beato, M. (1982) Monoclonal antibodies to the rat liver glucocorticoid receptor. *EMBO J.* **1**, 1467–1471.

75) Westphal, H.M., Mugele, K., Beato, M., and Gehring, U. (1984) Immunochemical characterization of wild-type and variant glucocorticoid receptors by monoclonal antibodies. *EMBO J.* **3**, 1493–1498.

76) Wrange, Ö., Carlstedt-Duke, J., and Gustafsson, J.-A. (1979) Purification of the glucocorticoid receptor from rat liver cytosol. *J. Biol. Chem.* **254**, 9284–9290.

77) Wrange, Ö., Okret, S., Radojcić, M., Carlstedt-Duke, J., and Gustafsson, J.-A. (1983) Characterization of the purified glucocorticoid receptor from rat liver cytosol. In "Steroid Hormone Receptors: Structure and Function," eds. Eriksson, H. and Gustafsson, J.-A., pp. 73–92, Elsevier Publ., Amsterdam.

78) Wrange, Ö., Okret, S., Radojcić, M., Carlstedt-Duke, J., and Gustafsson, J.-A. (1984) Characterization of the purified activated glucocorticoid receptor from rat liver cytosol. *J. Biol. Chem.* **259**, 4534–4541.

5

Glucocorticoid Hormone: Mechanisms of Action, ed. Sakamoto, Y. and Isohashi, F., pp. 119–135,
Japan Sci. Soc. Press, Tokyo/Springer-Verlag, Berlin (1986)

MECHANISMS OF GLUCOCORTICOID RESISTANCE IN LEUKEMIC CELLS

E. BRAD THOMPSON,*[1] YOSHIHIRO SAKAUE,*[2]
and JEFFREY M. HARMON*[3]

Department of Human Biological Chemistry and Genetics, The University of Texas Medical Branch, Galveston, Texas 77550, U.S.A.,[1] Osaka Infectious Disease Center, Department of Internal Medicine, Osaka 565, Japan,*[2] and Department of Pharmacology, Uniformed Services University of the Health Sciences, Bethesda, Maryland 20814, U.S.A.*[3]*

One major action of steroid hormones is to induce the synthesis of specific cell proteins. This was first shown to be so in rat liver and in liver-derived hepatoma cell lines in tissue culture (*27*). The use of cultured cells allowed the conclusions that the hormone acts directly on the cell induced, that it acts to increase synthesis of specific proteins and that it is required constantly to maintain induction (*25*). Many cellular actions of steroid hormones seem to be mediated through the interaction of the steroid with steroid group-specific hormone receptors. Most cells contain such glucocorticoid receptors, intracellular proteins of ~95,000 d. Active glucocorticoids, after entering cells, bind to this receptor, and this interaction seems essential for many of the known actions of the hormones. It is believed that the steroid-receptor complex directly interacts with other specific intracellular molecules to initiate some of these actions.

To induce some specific gene products it seems that the steroid-receptor complex binds to particular genomic DNA sequences, and that this interaction somehow provokes increased transcription of the gene (*18*, *21*, *30*). In other cases, glucocorticoids appear to cause a lengthening of the half-life of certain mRNAs, thereby causing accumulation of the mRNA pool and of its translated protein product (*6*). Before the steroid-receptor complex can bind DNA, however, it must undergo an "activation" step. Although the mechanism of this alteration is not understood, it is known that the change can be brought about by heat, by increased ionic strength, or by dilution of the cytosolic fraction in which the receptor-steroid complex has been prepared. We showed some time ago that the unactivated receptor complex can be separated conveniently from the activated by chromatography on DEAE- cellulose (*19*), and others found the same to be true for DEAE-Sephadex (*17*).

I. GLUCOCORTICOID RECEPTORS AND PROPERTIES OF THE ACTIVATED AND UNACTIVATED HORMONE-RECEPTOR COMPLEX

Figure 1 shows the pattern of HTC-cell and L-cell glucocorticoid-receptor complexes, before and after heat activation, chromatographed on DEAE-cellulose. As with rat liver, the initial complex, formed in hypo-osmotic buffer at 0–4°C, after being bound to the ion exchange resin, eluted in two peaks. The smaller fraction was eluted by a potassium phosphate concentration of ~0.05 M, the larger by ~0.2 M salt. After activation, whether by warming at 20°C or by dilution, the complex was found to elute almost entirely at the low salt position. The shift of receptor-bound steroid to the low salt eluting form was nearly quantitative. Each of the two forms of receptor-steroid complex rechromatographed in its original position when rerun on DEAE-cellulose (Fig. 2). Tests of binding showed that the low salt eluting form of the complex almost exclusively contained the DNA, chromatin, and nuclear binding capacity of the preparations (Table I, Fig. 3, ref. *19*). On the basis of such experiments, we concluded that it represented "activated" steroid-receptor complex. It is known that not only activation but also partial proteolysis can cause the steroid-receptor complex to convert to a form binding DNA with increased avidity. We therefore examined the size of the activated complex by sucrose gradient centrifugation. In low salt buffer the activated complex displayed a sedimentation coefficient of 4.6 S; in 0.3 M KCl, this complex ran with a coefficient of 3.7S. The unactivated complex (high salt peak from DEAE-cellulose) ran at 8S in low salt gradient

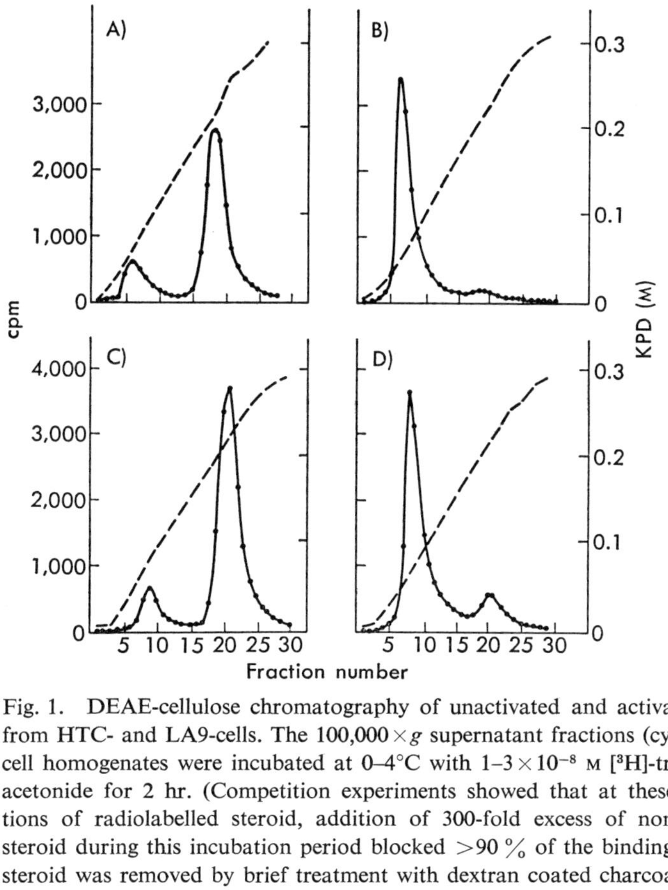

Fig. 1. DEAE-cellulose chromatography of unactivated and activated cytosols from HTC- and LA9-cells. The $100{,}000 \times g$ supernatant fractions (cytosols) from cell homogenates were incubated at 0–4°C with $1\text{–}3 \times 10^{-8}$ M [^{3}H]-triamcinolone acetonide for 2 hr. (Competition experiments showed that at these concentrations of radiolabelled steroid, addition of 300-fold excess of non-radioactive steroid during this incubation period blocked >90 % of the binding). Unbound steroid was removed by brief treatment with dextran coated charcoal. Half portions of each labelled cytosol were incubated simultaneously at 20°C and 2°C for 1 hr. Each was applied to a DEAE-cellulose column which was then thoroughly washed with 40 ml 5 mM potassium phosphate buffer pH 7.4 (KPD) at 0–4°C. Only ~4 % of total applied counts were so eluted. Bound macromolecules were then eluted by a linear 0.005 M to 0.3 M KPD gradient. Fractions were collected and their radioactivity estimated by liquid scintillation counting. Panel A, HTC-cell triamcinolone acetonide-labelled cytosol after the post-labelling incubation at 0°C; panel B, HTC cytosol after post-labelling incubation at 20°C; panel C, LA9 cell cytosol, 0°C; panel D, LA9 cytosol, 20°C. All chromatography done at 0–4°C; dotted lines, experimentally determined gradients.

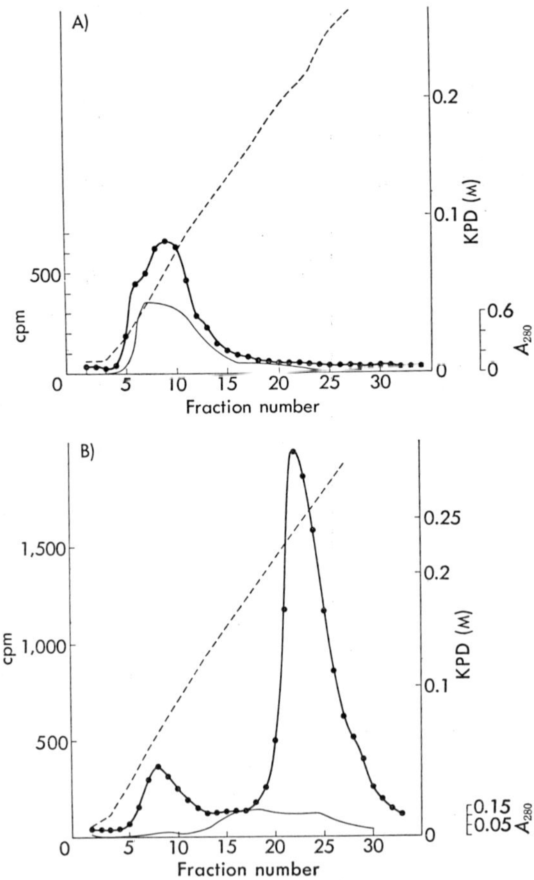

Fig. 2. Rechromatography of the low and high salt eluted peaks of glucocorticoid-receptor complexes. A rat liver cytosol was labelled with [^{3}H]-triamcinolone acetonide, applied to DEAE-cellulose, and eluted by a step gradient of 0.1 M; then, after washing, 0.3 M KPD. Each fraction was quickly desalted by gel filtration and rechromatographed on DEAE-cellulose with elution by a KPS gradient as in Fig. 1. Panel A, rechromatography of the peak originally eluted by 0.1 M KPD; panel B, rechromatography of the peak originally eluted by 0.3 M KPD; thin solid line, A_{280}.

TABLE I
DNA-cellulose Binding of the Receptor-Steroid Complex Eluted from DEAE-cellulose by Low Salt

	DNA-cellulose	Cellulose
Total applied	90,332 cpm (100%)	82,971 cpm (100%)
Did not bind	16,698 (18%)	68,455 (83%)
Bound, eluted with 0.5 M KCl	73,624 (82%)	14,516 (17%)

Rat liver cytosol was labelled at 0–4°C with [^{3}H]-triamcinolone acetonide, 1.6×10^{-8} M for 2 hr. Unbound steroid was removed by treating with dextran coated charcoal. The labelled cytosol was applied to a 3 ml column of DEAE-cellulose. After the column had been washed extensively with 0.005 M potassium phosphate buffer (KPB), pH 7.4, the bound radioactivity was eluted stepwise, first with 0.1 M KPB, and then—after further washing —with 0.3 M KPB. The low salt eluted fraction was divided and applied to either a DNA-cellulose column or a control cellulose column. The flow-through fractions and those obtained by washing the columns with 0.01 M Tris, 1.5 mM ethylenediamine tetraacetic acid, 0.5 mM dithiothreitol, pH 7.4 (TED) were pooled and counted as "did not bind." Bound counts were eluted with TED+0.5 M KCl. Similar experiments with the labelled cytosol eluted from DEAE-cellulose by 0.3 M potassium phosphate, after lowering its salt content, showed that it had but little capacity to bind to DNA.

and 3.7S in high salt gradients. Addition of proteinase inhibitors leupeptin, elastinal, antipain, phosphoramidon (46 μg/ml each), pepstatin and chymostatin (10-fold dilution of saturated solution), or trypsin inhibitor at 0.28 mg/ml failed to prevent activation, as measured by conversion of the high salt to the low salt eluting peak from DEAE-cellulose. Sizing by gel filtration on Biogel A-0.5m also indicated the activated, low salt eluting peak to be undegraded. Therefore, we do not believe that proteolysis in this system accounts for the shift in DEAE-elution pattern and affinity for DNA seen after activation.

In summary, two distinct forms of the undegraded glucocorticoid-receptor complex can be separated (Table II). The "unactivated" form is the main one produced when corticosteroids are allowed to bind to receptors at 0–4°C in cell cytosolic fractions prepared in the usual dilute buffers. "Unactivated" receptor binds to DEAE-cellulose relatively tightly, and is eluted by about 0.2 M potassium phosphate. The "activated" receptor-steroid complex is formed by a nearly quantitative shift from the "unactivated" form, following dilution of the cytosol at continued low temperature, shift-up to 20°C or increased salt concentration. The "activated" complex binds DEAE-cellulose also but less firmly, and is eluted by ~0.05 M phosphate.

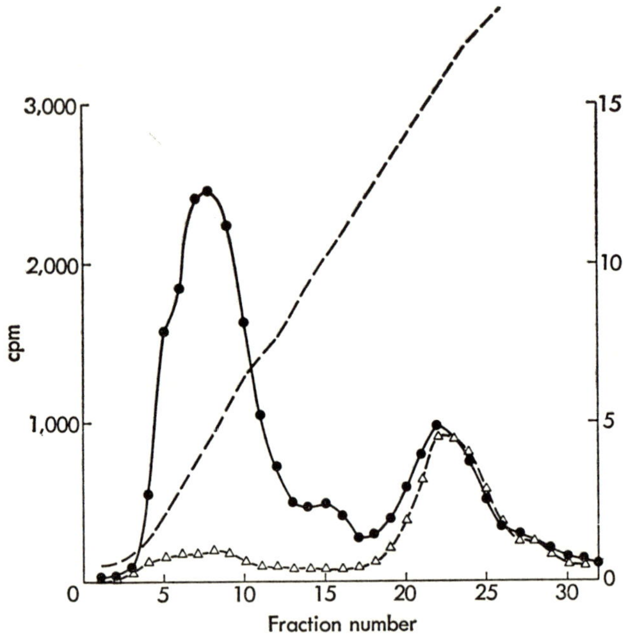

Fig. 3. DEAE-cellulose chromatography of post-chromatin supernatant of heat-activated cytosol. Rat liver cytosol was incubated with [^{3}H]-triamcinolone acetonide at 0°C for 2 hr, and unbound steroid was removed as described in Fig. 1. The labeled cytosol was incubated at 20°C for 30 min. Chromatin prepared from rat liver as described elsewhere (*19*) was suspended in 5 mM KPD buffer and precipitated by centrifugation at 12,000 $\times g$ for 10 min. An aliquot of the activated cytosol was mixed with the pelleted chromatin and incubated for 1.5 hr at 0°C. The mixture was agitated every 3–5 min. After the incubation, the tubes were centrifuged at 12,000 $\times g$ for 10 min, and the supernatant was chromatographed on DEAE-cellulose as described in the legend to Fig. 1. For comparison, untreated activated cytosol equal in volume with the chromatin-treated cytosol was also chromatographed. Activated cytosol not treated with chromatin (●); activated cytosol treated with chromatin (△); conductivity (salt concentration) (- - - -).

This form of the complex binds to DNA, chromatin, and nuclei. Do these receptor complex properties in broken cell systems apply *in vivo?* There is some evidence that they do. Most obvious is the fact that as prepared, cytosolic fractions from liver of intact animals contain mostly the "unactivated" form, as do similar fractions from L (transformed mouse fibroblasts) or

TABLE II
Properties of "Activated" and "Unactivated" Glucocorticoid-Receptor Complexes

	Activated	Unactivated
Relative quantity in dilute homogenates	Less	More
Sucrose velocity sedimentation gradient properties		
In low salt	4.6S	8S[a]
In high salt	3.7S	3.7S
Eluted from DEAE-cellulose by	0.05 M PO_4^{2-}	0.2 M PO_4^{2-}
Formed by	Binding steroid at 4°C, then elevating temperature to 20°C, salt to 0.3–0.4 M or by dilution of cytosol	Binding steroid at 0–4°C
Binds to		
DNA	+	−
Chromatin	+	−
Nuclei	+	−

[a] Data for the 0.2 M PO_4^{2-} peak off DEAE-cellulose, desalted and run on sucrose velocity gradient. Unfractionated cytosol gives a (presumably unactivated) peak at ~10S. Data taken from refs. *9*, *19*, *20*, *31* and unpublished results.

HTC (transformed rat hepatocytes) cells in culture. Secondly, in normal living thymocytes, timed studies following the appearance of the two forms after addition of steroids support the view that the transition actually occurs *in vivo* (*13*, *14*). Thus the general properties of activated and unactivated steroid receptors seem to apply to living rodent thymocytes. These properties also apply to human leukemic lymphoblasts, and form the basis for a new class of corticoid resistance, as will be seen below.

II. CEM-CELLS AND THE PROPERTIES OF GLUCOCORTICOID RECEPTORS IN HUMAN LEUKEMIC BLASTS

CEM-cells are a cultured line of human leukemic lymphoblasts, grown from the peripheral blood of a female patient with acute lymphoblastic leukemia (*3*). Among several cell lines that we originally screened, we found CEM to be one of the more steroid-sensitive. We therefore cloned it, rescreened the clones and chose clone CEM C7 to be our prototype glucocorticoid-sensitive line (*16*). We have shown these cells to have three responses to glucocorticoids: induction of glutamine synthetase, irreversible cessation of growth in the G_1

TABLE III
Loss of Viability of CEM C7 Cells in Continuous 10^{-6} M Dexamethasone in Growth Medium

Relative cloning efficiency (%)	Time after hormone added (hr)
100	0
100	12
100	18
80	24
50	30
23	36
13	42
8	48
2	60

phase of the cell cycle, and cell lysis (*8*, *11*, *16*). The cells contain glucocorticoid-specific receptors, and the concentrations of active hormone that occupy these receptors are those that evoke the biological responses (*8*, *11*, *16*). Continuous occupancy of receptors is required to maintain glutamine synthetase induction, and if the hormone is removed from the cells, the enzyme promptly returns to deinduced levels (*11*). Continuous occupancy of receptors also is required to elicit the growth-arresting and lytic responses, but since these are in themselves irreversible, removal of steroid cannot stop them in those cells already so committed. The processes that lead to the final effects of G_1 arrest and lysis, however, are slow, so that they may be interrupted by prematurely washing out the hormone from the cells. For example, G_1 arrest/lysis only begins to be seen in C7 after ~24 hr steroid exposure. Removing hormone prior to that time leaves the cells fully capable of growth. At times $\geq$in 24 hr in steroid, more and more cells arrest and die; but still many will survive if steroid is removed (*8*). For maximum cell kill, full receptor-occupying concentrations of hormone, continuously, are required for several days (Table III).

The cells of the CEM line have a modal chromosomal number of 47, with a pericentric inversion and partial loss of chromosome 9 plus trisomy 20 (*12*). Surface markers suggest that they are of T-cell origin. They make no interleukin 2, nor do they require it for growth. The physical properties of the glucocorticoid receptors of these cells have been explored. These are

TABLE IV
Some Properties of the Human Glucocorticoid Receptor from CEM C7-Cells

Affinity for dexamethasone (whole-cell binding)	$K_d=2\times10^{-8}$, single class of affinity sites
Velocity gradient centrifugation	
low salt	~10S
high salt	~4S
Gel filtration, Stokes radius	56 Å
Activated complex eluted from DNA-cellulose by	0.1 M PO_4^{2-}
Covalently labelled receptor:	
Mr by SDS-PAGE[a]	~95,000
Eluted from RP-HPLC[b] by	55% acetonitrile
Elution from DEAE-cellulose	
Unactivated	0.2 M PO_4^{2-}
Activated	0.05 M PO_4^{2-}

[a] Sodium dodecyl sulfate-polyacrylamide gel electrophoresis. [b] Reverse phase-high performance liquid chromatography.

given in Table IV. The receptors bind [^{3}H]-dexamethasone, in a whole-cell binding assay, with an apparent equilibrium dissociation constant of 2×10^{-8} M. Scatchard plots of binding data are consistent with a single class of receptor sites. Labelled crude receptor, unactivated, yields a sedimentation constant of ~10S on sucrose gradients and ~4S on high salt gradients. The Stokes radius of the molecule is ~56 Å by gel filtration. When activated, it binds to DNA-cellulose, from which it is eluted by 0.1 M potassium phosphate. Dexamethasone mesylate, ^{3}H-labeled, covalently attaches to the glucocorticoid receptor, allowing subsequent electrophoresis on polyacrylamide gels under denaturing conditions. Doing so shows that the undegraded receptor is a single-sized protein of ~95,000 d. This is an Mr similar to that which can be calculated from the gel filtration plus sucrose gradient data. In addition, the 95,000 d, covalently-labelled receptor elutes from reverse-phase high performance liquid chromatography (HPLC) columns at a concentration of ~55 % acetonitrile (Thompson, unpublished results). As is easily seen, the human receptor closely resembles that from rodent cells. Finally, it also does so in showing similar differential binding to DEAE-cellulose. The unactivated CEM C7 receptor eluted from such columns as ~0.2 M potassium phosphate is reached in a linear gradient. The activated form binds only weakly, and is eluted by 0.05 M phosphate.

III. THE GENETICS OF GLUCOCORTICOID RESISTANCE IN CEM-CELLS

Receptor Mutants

The strong selective pressure exerted on CEM C7-cells by the lethal effects of glucocorticoids was taken advantage of to obtain a collection of resistant clones of cells (*10*). To increase the probability that each of these arose independently and at the same time to provide information about the rate of development of resistance, 4 Luria-Delbrück fluctuation experiments were carried out. From the data obtained a rate of generation of resistance was calculated to be $\sim 10^{-5}$ per cell generation. Some 50-odd dexamethasone resistant clones were screened in preliminary fashion by quantitating their binding sites for [^{3}H]-dexamethasone. Easily measured but lower than wild-type numbers of sites were found in all these spontaneously resistant clones. Two tests of function were used to screen the receptors: "nuclear translocation" and glutamine synthetase induction. "Nuclear translocation" refers to the amount of labelled steroid-receptor complex found in the cytosolic *vs.* the crude nuclear fraction from cells exposed to [^{3}H]-dexamethasone at 37°C. Under such conditions in CEM C7, our prototype dexamethasone-sensitive cells, about 40 % of the labelled complex is found to be associated with the nuclear fraction. In all the spontaneously resistant clones, this percentage was decreased. In the second test, none of these clones showed induction of glutamine synthetase by dexamethasone. These results suggested an abnormality of receptors or of the cell responses to receptors; so three clones were selected for more detailed examination. More detailed binding studies with [^{3}H]-dexamethasone confirmed that all three had reduced numbers of specific binding sites. Binding to these sites produced data which could be best plotted on linear Scatchard plots, with slopes suggesting 2–7-fold less affinity for this steroid than CEM C7 receptors had. The resistant cells' receptors were then labelled covalently with [^{3}H]-dexamethasone mesylate (*24*), precipitated with our antiserum to the human receptor, and compared on denaturing polyacrylamide gel electrophoresis (*7*, *29*). By this test their size was identical to that of CEM C7 receptor run on the same gel. Mixing experiments gave no evidence for factors in the resistant cells that interfere with wild-type receptors' binding of steroid.

When the three resistant clones' receptors were examined on DEAE-cellulose, however, an abnormality was found (*9*, *20*). In the unactivated state, they were eluted from the ion exchange column in a fashion identical to wild-type, by ~0.2 M potassium phosphate. But unlike normal receptors when exposed to 20°C to cause activation, they lost their ability to retain the bound steroid; so instead of shifting to the low salt eluting position, they simply released the steroid. Adding molybdate salts to steroid hormone receptors is known to stabilize them in their unactivated state. When this is done to dexamethasone-labelled CEM C7 cytosol, heating to 20°C does not cause a shift to the low salt elutable, "activated" form. The same was found true for two of the three clones of resistant cells. The third clone's receptors were not protected by molybdate, but lost their steroid-binding capacity at 20°C anyway. On the basis of these data we have proposed that these cells have a previously undescribed phenotype of receptor-based resistance, which we call activation-labile or r act^l. Clearly there are two subclasses, molybdate-sensitive (Mo^s) and molybdate-resistant (Mo^r), hinting that perhaps two mutations gave rise to quite similar phenotypes. The r act^l receptors do not seem inherently less heat stable than normal receptors when they have not bound steroid. If cytosols without steroid added are held at 20°C, with samples removed at intervals and tested for residual ability to bind glucocorticoid, all three mutants and CEM C7 have the same rate of loss of binding ability. We have recently carried out somatic cell hybridizations of these cells to test for dominance/recessive properties and for complementation. The results of these experiments will follow the description of a second, different type of dexamethasone-resistant mutant.

By far the predominant phenotype observed when rodent cells were selected for glucocorticoid resistance was that of receptor deficiency, *i.e.*, simple loss of part or all the binding sites (*1*, *22*, *23*). In CEM-cells, we have yet to find this phenotype in spontaneous mutants. However, when CEM C7-cells were first treated with either frame-shift or point mutagens, then selected in 10^{-6} M dexamethasone, most of the resistant clones obtained had $\leq 10\,\%$ of wild-type binding. These receptor-deficient or r^- cells show this low binding in whole-cell assays. Cytosolic preparations often show little binding, suggesting that the residual sites are unstable. Mixing experiments for the clones tested to date give no evidence for injurious non-receptor components in these cells. Glutamine synthetase is not induced in them by dexamethasone. When the cells of one such clone, ICR 27, were labelled with [^{3}H]-dexamethasone mesylate, and the receptors immunoprecipitated and electrophoresed as described above for the r act^l clones, a trace quantity

TABLE V
Somatic Cell Hybrid Analysis of r act^1 and r$^-$ Mutants

Parents	Receptor phenotype	Selective marker	Responses to dex: Growth	Responses to dex: Glutamine synthetase
C7	r$^+$	HGPRT$^-$	Inhibited	Induced
C7GS$^-$	r$^+$	TK$^-$	Inhibited	Not induced
3R7	r act^{1}Mo^s	HGPRT$^-$	Unaffected	Not induced
3R43	r act^{1}Mo^r	HGPRT$^-$	Unaffected	Not induced
ICR27	r$^-$	TK$^-$	Unaffected	Not induced
Hybrids				
C7×ICR27	(r$^+$×r$^-$)		Inhibited	Induced
C7GS$^-$×3R43	(r$^+$×r act^{1}Mo^r)		Inhibited	Induced
C7GS$^-$×3R7	(r$^+$×r act^{1}Mo^s)		Inhibited	Induced
ICR27×3R43	(r$^-$×r act^{1}Mo^r)		Unaffected	Not induced
ICR27×3R7	(r$^-$×r act^{1}Mo^s)		Unaffected	Not induced

of ~95,000d receptor was found (*29*). Thus it seems that this clone, and by inference the others with reduced binding sites, simply have too few receptors to respond properly.

Selective markers were placed in CEM C7, both sub-types of r act^1 mutants, and the r$^-$ mutant ICR 27. By standard methods, somatic cell hybrids were produced, crossing wild-type cells with each type of mutant and the mutants with each other. Hybrid formation was confirmed by chromosome counts as well as growth in selective medium. In all crosses involving mutant×parent, receptor site content was roughly additive. Glutamine synthetase induction and growth inhibition both were observed after addition of dexamethasone. Thus no evidence was found for negative cooperativity or for trans-active elements blocking the steroid responsiveness contributed by the normal genome. Nor was there evidence for positive cooperativity between the three classes of resistant cells. Again, total receptor sites were roughly additive, but neither functional response was restored (Table V). Of note were the results of the crosses between the CEM C7 thymidine kinase deficient (TK$^-$) parent and the r act^1 mutants. This C7 r$^+$ parent, selected for resistance to BUDR, was found to have lost glutamine synthetase (GS$^-$) inducibility but was still fully sensitive to the growth inhibitory effects of glucocorticoids. (Such non-coordinate loss of some parts of an array of inducible functions had previously been observed in clones of HTC-cells, a glucocorticoid sensitive rat hepatoma cell line, ref *26*). Thus the C7 (TK^- r^+ GS^-)$\times$r act^1 crosses, which resulted in hybrids inducible for glutamine synthetase, suggest that the receptor supplied by the C7 cells was acting to induce the GS locus of the r act^1 cells. This is additional evidence that the mutation in these cells was limited to the steroid receptors and was not one generally affecting steroid responses in the mutants.

These somatic cell mutants demonstrate several important points. First, they prove that glucocorticoid receptors are essential components of the mechanism through which the hormones directly kill certain human leukemia cells. Second, the CEM C7 $r^+GS^-TK^-$ clone shows that glutamine synthetase induction is not an essential component of that process. The r act^1 mutants indicate that receptor-steroid complex activation is essential for enzyme induction and for cell kill; they also indicate one possible mechanism by which human leukemias could become steroid-resistant without losing all hormone binding, a commonly found circumstance. It is interesting also that recently a patient with end-organ cortisol resistance was described who seems to have activation-labile glucocorticoid receptors (*15*). The somatic cell hybrids we have studied also suggest that the active steroid-receptor complex is a single class of molecule. It is generally found that when unactivated, glucocorticoid receptors are prepared in dilute buffers in crude cytosolic fractions, the steroid-receptor complexes are found in large aggregates. Note above, for example, the 10S sedimentation constant obtained for such complexes both from rat and human cells. Whether such aggregates exist in the cell is unknown, as is their precise composition; but their occurrence raises the possibility of dissimilar steroid-binding receptor subunits participating in a critical step of the hormone effects. The hybrid cell data do not support this possibility, since such a model would predict negative complementation between normal and abnormal receptors, as in the $r^+ \times$ r act^1 crosses. Instead, we observed co-dominance, as though there were no receptor interactions.

IV. LYSIS-DEFECTIVE VARIANTS

It has long been predicted that there are specific steps "beyond the receptor" in glucocorticoid-evoked cell lysis, but discovering those steps has been difficult. Recently reported however, were experiments of interest relative to this problem. In the SAK8 mouse thymus cell line, the cells of which contain glucocorticoid receptors and display several inducible functions, glucocorticoids were unable to affect cell growth. However, if these cells were treated with 5-azacytidine, a compound known to interfere with DNA methylation, they became sensitive to lysis by corticosteroids (*4*, *5*). We have isolated from CEM-cells an analogous clone, CEM C1 (*31*). This clone was obtained from the original CEM line by random cloning without glucocorticoid selection. It grows equally well in the presence or absence of 10^{-6} M dexamethasone, has receptors identical in every respect so far tested to those of CEM C7-cells, and contains inducible glutamine synthetase. Thus it can be described as lysis defective or ly^-, and is a candidate for having a lesion "beyond the receptor" which blocks the lytic response. In collaboration with Smith and Bourgeois, we have treated these cells with 5-aza-2′-deoxycytidine, to block DNA methylation. When the cells so treated were recloned, a number of the resultant clones were found to be fully sensitive to growth arrest and lysis by dexamethasone (*28*). Although the precise mechanism of this change from ly^- to ly^+ phenotype is unknown, it may be that the demethylation evoked by the cytidine analog has allowed re-expression of some gene(s) critical for the lytic response. It is known that cellular DNA methylation occurs on cytidine residues during S phase as the result of the action of a specific methylase. Methylation of certain loci has been correlated with the loss of expression of a large number of genes (*2*). Thus, by detailed analysis of CEM C1-cells and their azacytidine-induced revertants, we may obtain new insights into the critically important steps by which glucocorticoids stop the growth of leukemic cells.

SUMMARY

We have described a method of distinguishing the activated from the unactivated form of the glucocorticoid-receptor complex and shown that these two forms can be separated in rat liver and HTC-cells, mouse L-cells, and human CEM-cells. The latter are a line of human leukemic lymphoblasts, normally lysed by active glucocorticoids. These cells contain glucocorticoid receptors, and these must be continually occupied for $\geq$24 hr before the process of cell lysis begins to be apparent. Glutamine synthetase is also induced by such hormones in these cells. From the sensitive clone CEM C7 we have obtained r^- and r act^1 mutant subclones, all of which are resistant to the effects of dexamethasone. Somatic cell hybrids among these clones indicate they represent classes of independent receptor defects. These clones establish the essential participation of the glucocorticoid receptor in corticoid-induced lysis of human leukemic cells. A different sort of resistant clone, CEM C1, contains normal receptors but is lysis defective (r^+ ly^-). Treatment of these cells with 5-aza-2′-deoxycytidine can restore the lytic response.

Acknowledgments
This work was supported in part by grant CA 32226 from the National Cancer Institute to J. M. H.

REFERENCES

1) Bourgeois, S., Newby, R.F., and Huet, M. (1978) Glucocorticoid resistance in murine lymphoma and thymona lines. *Cancer Res.* **38**, 4279–4284.
2) Doerfler, W. (1983) DNA methylation and gene activity. *Annu. Rev. Biochem.* **52**, 93–124.
3) Foley, G.E., Lazarus, H., Farber, S., Uzman, B.F., Boone, B.A., and McCarthy, R.E. (1965) Continuous culture of human lymphoblasts from peripheral blood of a child with acute leukemia. *Cancer* **18**, 522–529.
4) Gasson, J.C. and Bourgeois, A. (1983) A new determinant of glucocorticoid sensitivity in lymphoid cell lines. *J. Cell Biol.* **96**, 409–415.
5) Gasson, J.C., Ryden, T., and Bourgeois, S. (1983) Role of *de novo* DNA methylation in the glucocorticoid resistance of a T-lymphoid cell line. *Nature* **302**, 621–623.
6) Guyette, W.A., Matusik, R.J., and Rosen, J.M. (1979) Prolactin-mediated transcriptional and post-transcriptional control of casein gene expression. *Cell* **17**, 1013–1023.
7) Harmon, J.M., Eisen, H.J., Brower, S.T., Simons, S.S., Jr., Langley, C.L., and Thompson, E.B. (1984) Identification of human leukemic glucocorticoid receptors using affinity labeling and anti-human glucocorticoid receptor antibodies. *Cancer Res.* **44**, 4540–4547.
8) Harmon, J.M., Norman, M.R., Fowlkes, B.J., and Thompson, E.B. (1979) Dexamethasone induces irreversible G, arrest and death of a human lymphoid cell line. *J. Cell. Physiol.* **98**, 267–278.
9) Harmon, J.M., Schmidt, T.J., and Thompson, E.B. (1984) Molybdate-sensitive and molybdate-resistant activation-labile glucocorticoid-receptor mutants of the human lymphoid cell line CEM-C7. *J. Steroid Biochem.* **21**, 227–236.
10) Harmon, J.M. and Thompson, E.B. (1981) Isolation and characterization of

dexamethasone-resistant mutants from human lymphoid cell line CEM-C7. *Mol. Cell. Biol.* **1**, 512–521.

11) Harmon, J.M. and Thompson, E.B. (1982) Glutamine synthetase induction by glucocorticoids in the glucocorticoid-sensitive human leukemic cell line CEM-C7. *J. Cell. Physiol.* **110**, 155–160.

12) Moore, D.E., Weise, K., Zawydiwski, R., and Thompson, E.B. (1985) The karyotype of the glucocorticoid-sensitive lymphoblastic human T cell line CCRD-CEM shows a unique deleted and inverted chromosome #9. *Cancer Genet. Cytogenet.*, in press.

13) Munck, A. and Foley, R. (1979) Activation of steroid hormone-receptor complexes in intact target cells in physiological conditions. *Nature* **278**, 752–754.

14) Munck, A. and Foley, R. (1980) Activated and non-activated glucocorticoid receptor complexes in rat thymus cells: kinetics of formation and relation to steroid structure. *J. Steroid Biochem.* **12**, 225–230.

15) Nawata, H., Sekiya, K., Higuchi, K., Yanase, T., Kata, K., and Ibayashi, H. (1984) Thermolabile and activation-labile glucocorticoid receptor of cultured skin fibroblast in the patient of glucocorticoid resistance syndrome. In "Abstracts, 7th International Congress of Endocrinology," p. 1063, Excerpta Medica, Amsterdam, Oxford-Princeton.

16) Norman, M.R. and Thompson, E.B. (1977) Characterization of a glucocorticoid-sensitive human lymphoid cell line. *Cancer Res.* **37**, 3785–3791.

17) Parchman, L.G. and Litwack, G. (1977) Resolution of activated and unactivated forms of the glucocorticoid receptor from rat liver. *Arch. Biochem. Biophys.* **183**, 374–382.

18) Payvar, F. and Wrange, O. (1983) Relative selectivities and efficiencies of DNA binding by purified intact and protease-cleaved glucocorticoid receptor. In "Steroid Hormone Receptors: Structure and Function," eds. Eriksson, H. and Gustafsson, J.-A., pp. 267–284, Elsevier, Amsterdam.

19) Sakaue, Y. and Thompson, E.B. (1977) Characterization of two forms of glucocorticoid hormone-receptor complex separated by DEAE-cellulose column chromatography. *Biochem. Biophys. Res. Commun.* **7**, 533–541.

20) Schmidt, T.J., Harmon, J.M., and Thompson, E.B. (1980) 'Activation-labile' glucocorticoid-receptor complexes of a steroid-resistant variant on CEM-C7 human lymphoid cells. *Nature* **286**, 507–510.

21) Schrader, W.T., Compton, J.G., and O'Malley, B.W. (1983) Progesterone receptor binding to the ovalbumin gene. In "Steroid Hormone Receptors: Structure and Function," eds. Eriksson, H. and Gustafsson, J.-A., pp. 247–265, Elsevier, Amsterdam.

22) Sibley, C.H. and Tomkins, G.M. (1974) Isolation of lymphoma cell variants resistant to killing by glucocorticoids. *Cell* **2**, 213–220.

23) Sibley, C.H. and Tomkins, G.M. (1974) Mechanisms of steroid resistance. *Cell* **2**, 221–227.
24) Simons, S.S., Jr. and Thompson, E.B. (1981) Dexamethasone-21-mesylate: an affinity label of glucocorticoid receptors from rat hepatoma tissue culture cells. *Proc. Natl. Acad. Sci. U.S.A.* **78**, 3541–3545.
25) Thompson, E.B. (1979) Glucocorticoid induction of tyrosine aminotransferase in cultured cells. In "Monographs on Endocrinology, Vol. 12, Glucocorticoid Hormone Action," eds. Baxter, J.D. and Rousseau, G.G., pp. 203–217, Springer-Verlag, Berlin.
26) Thompson, E.B., Granner, D.K., Gelehrter, T., Erickson, J., and Hager, G.L. (1979) Unlinked control of multiple glucocorticoid-induced processes in HTC cells. *Mol. Cell. Endocrinol.* **15**, 135–150.
27) Thompson, E.B. and Lippman, M.E. (1974) Mechanism of action of glucocorticoids. *Metabolism* **23**, 159–202.
28) Thompson, E.B., Smith, J.R., Bourgeois, S., and Harmon, J.M. (1985) Glucocorticoid receptors and human leucopathic diseases. *Klin. Wochenschr.* **63**, 689–698.
29) Thompson, E.B., Zawydiwski, R., Brower, S.T., Eisen, H.J., Simons, S.S., Schmidt, T.J., Schlechte, J.A., Moore, D.E., Norman, M.R., and Harmon, J.M. (1983) Properties and function of human glucocorticoid receptors in steroid-sensitive and -resistant leukemic cells. In "Steroid Hormone Receptors: Structure and Function," eds. Eriksson, H. and Gustafsson, J.-A., pp. 171–194, Elsevier, Amsterdam.
30) Yamamoto, K.R. (1983) On steroid receptor regulation of gene expression and the evolution of hormone-controlled gene networks. In "Steroid Hormone Receptors: Structure and Function," eds. Eriksson, H. and Gustafsson, J.-A., pp. 285–306, Elsevier, Amsterdam.
31) Zawydiwski, R., Harmon, J.M., and Thompson, E.B. (1983) Glucocorticoid-resistant human acute lymphoblastic leukemic cell line with functional receptor. *Cancer Res.* **43**, 3865–3873.

6

Glucocorticoid Hormone: Mechanisms of Action, ed. Sakamoto, Y. and Isohashi, F., pp. 139–153,
Japan Sci. Soc. Press, Tokyo/Springer-Verlag, Berlin (1986)

CORRELATION BETWEEN HORMONE-MEDIATED MESSENGER RNA INDUCTION AND NUCLEAR HISTONE ACETYLATION

MASAAKI HIROSE

The Research Institute for Food Science, Kyoto University, Uji 611, Japan

The ultimate responses to hormones that cause growth or differentiation of their target tissues are believed to be the changes in the complements of specific proteins. The mammary gland is one of the excellent model systems for investigating the mechanism of hormone action, since milk proteins such as casein are abundantly synthesized in this tissue under hormone control. Milk protein synthesis is induced by the interaction of prolactin and glucocorticoid. In contrast, progestin has the potential to prevent the milk protein synthesis during pregnancy.

Mammary hormone-binding receptors have been extensively studied to ascertain the specific roles of the steroid and peptide hormones on milk protein induction. The presence of specific prolactin-binding receptors on the surface of mammary cells has been documented (*2*). Cytoplasmic glucocorticoid receptors have been characterized with the mammary glands of rodents (*14*, *15*, *19*, *25*). Our detailed studies have revealed that a synthetic progestin, R5020, competitively binds to glucocorticoid receptor binding sites with an affinity similar to dexamethasone (*7*, *15*), suggesting that progestin strongly competes with glucocorticoid on receptor binding sites and blocks its stimulation effect. The biochemical processes involved in the milk protein induction following the binding of prolactin and glucocorticoid to their specific receptors, however, have not been established.

This report presents the results of our studies on the nuclear processes of hormone action involved in milk protein induction in the rat mammary gland.

I. NUCLEAR BINDINGS OF GLUCOCORTICOID-RECEPTOR COMPLEXES

Nuclear bindings of cytoplasmic glucocorticoid-receptor complexes of mammary glands were investigated under cell-free conditions (*8*). These complexes were prepared by incubating cytosol from lactating rat mammary glands with radiolabeled dexamethasone (native receptor). The native receptor complexes were exposed to a high concentration of ammonium sulfate and salt was removed by passing them through a Sephadex G-25 column ($(NH_4)SO_4$-treated receptor). As shown in Fig. 1, on DNA-cellulose column chromatography the $(NH_4)_2SO_4$-treated receptor complexes separated into two peaks which consisted of DNA-cellulose-bound and non-DNA-cellulose-bound fractions. When the sample containing the non-DNA-cellulose-bound receptor complexes was subjected to re-chromatography on a DNA-cellulose column, more than 95% of the receptor activity was recovered in wash-through fractions. On the contrary, in the re-chromatography of the DNA-cellulose-bound receptor complexes, 86% of the receptor activity was recovered in DNA-cellulose-bound fractions. Therefore, there exist at least two receptor forms in the $(NH_4)_2SO_4$-treated receptor complexes: one is bound to DNA-cellulose (form I receptor), and the other is non-bound to DNA-cellulose (form II receptor).

Figure 1 also shows that when the native receptor was applied to a DNA-cellulose column, receptor activity was detected only in wash-through fractions. The same was true when [^{3}H]-dexamethasone-receptor complexes treated with ammonium sulfate in the presence of sodium molybdate were applied to a DNA-cellulose column. Since molybdate inhibits steroid receptor transformation into activated states (*11*, *28*), the form II receptor appeared to be in a non-activated state using the criterion of DNA-cellulose column chromatography.

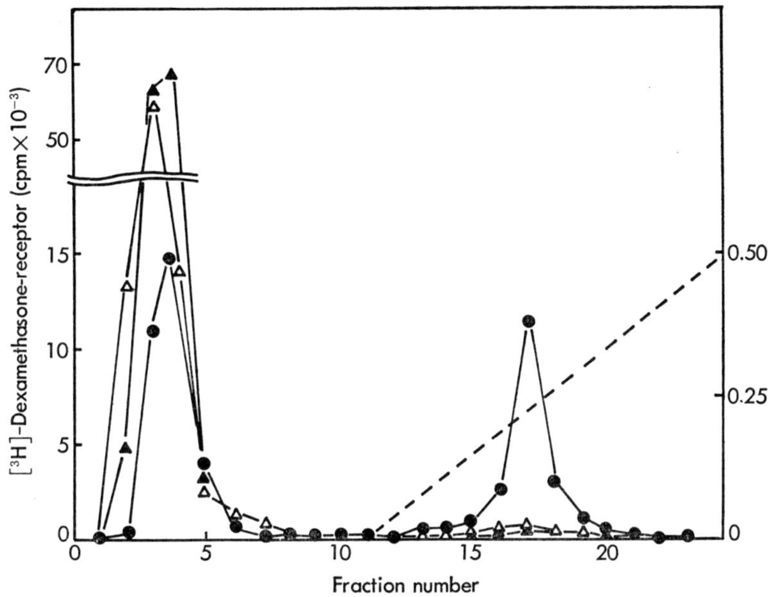

Fig. 1. DNA-cellulose column chromatography of mammary glucocorticoid-receptor complexes. Cytosol labeled with [^{3}H]-dexamethasone (the native receptor) (▲), the [^{3}H]-dexamethasone-receptor complexes treated with ammonium sulfate in the presence of 10 mM Na_2MoO_4 (△), or the [^{3}H]-dexamethasone-receptor complexes treated with ammonium sulfate (the $(NH_4)_2$-treated receptor) (●) were applied to a DNA-cellulose column (0.7×2.0 cm, 0.7 mg DNA) equilibrated with Tris/EDTA buffer. The column was washed with the same buffer, then developed with a linear 0–0.5 M NaCl gradient in the same buffer. Aliquots of the fractions (each fraction, 0.8 ml) were withdrawn, and activity of the [^{3}H]-dexamethasone-receptor complex was assayed by the dextran-coated charcoal technique.

In several glucocorticoid receptor systems, activated receptors were eluted from a DEAE-cellulose column at a lower salt concentration than non-activated receptors *(24)*. A test was made of whether the $(NH_4)_2SO_4$-treated receptor complexes are also separated into two peaks with similar quantities in DEAE-cellulose column chromatography. However, about 90% of these receptor complexes were eluted at a low salt concentration of 0.05–0.15 M NaCl, whereas almost all the native receptor complexes were eluted at a higher salt concentration (0.15–0.25 M NaCl). Therefore, by the criterion of DEAE-cellulose column chromatography, most of the glucocorticoid-receptor complexes in the $(NH_4)_2SO_4$-treated receptor sample appeared to be activated.

The question was raised whether the non-DNA-cellulose-bound receptor (the form II receptor) is in an activated state. It has been shown that glucocorticoid-receptor complex bindings to DNA-cellulose and nuclei are different, and that nuclei should be used for accurate measurements of activated steroid-receptor complexes (*12*). Molybdate inhibits temperature- or salt-dependent steroid receptor transformation into activated states, but it is not inhibitory against once activated steroid-receptor complexes (*11*, *28*). To answer the question of whether the form II receptor is activated, its nuclear binding abilities were tested in the presence or absence of molybdate. Both the form II and the form I receptor exhibited the ability to bind to mammary gland nuclei. The presence of molybdate did not significantly inhibit the nuclear binding of these receptor forms. However, the binding ability of [^{3}H]-dexamethasone-receptor complexes which had been treated with ammonium sulfate in the presence of molybdate was very small. These data thus indicate qualitatively that both the form I and form II receptors are in an activated state.

To estimate the concentration of activated receptor complexes quantitatively, a fixed concentration of the two forms of the mammary glucocorticoid-receptor complexes was titrated with increasing concentrations of nuclei; titration profiles were not greatly different between the two forms. More than 60% of the form II receptor was bound by a high concentration of nuclei (0.9 mg/ml DNA), while 73% of the form I receptor was bound by the same concentration. Therefore, most, if not all, of the form II receptor appears to be in an activated state.

Nuclear chromatin has been shown to be one of the main nuclear acceptor sites for steroid-receptor complexes (*27*); however, it has been argued that other classes of sites are also involved in the nuclear binding of steroid-receptor complexes (*26*). To test the possibility that they bind to distinct nuclear components, form I and form II receptors were tested in two different types of experiments for their ability to bind to nuclear chromatin.

In one experiment, glucocorticoid-receptor complexes were incubated with nuclei and then the chromatin fraction was isolated. If the form I and form II receptors had recognized different nuclear components, the amounts of dexamethasone-receptor complexes bound to chromatin would have been vastly different between the two; Table I shows that this was not the case.

TABLE I
Binding of the Two Receptor Forms to Nuclei and Chromatin

Receptor	Glucocorticoid-receptor binding (fmol/mg DNA)	
	Nuclei	Chromatin
Experiment A		
Form I receptor	972±86	584±40
Form II receptor	855±72	567±34
Experiment B		
Form I receptor	533±30	369±74
Form II receptor	343±23	338±38

In experiment A, 4.0 pmol of [^{3}H]-dexamethasone-receptor complexes (the form I or form II receptor) were incubated with mammary gland nuclei (2.5 mg DNA) at 0°C for 90 min in a total volume of 2.5 ml. Aliquots of the samples were withdrawn and the nuclear bindings of glucocorticoid-receptor complexes were determined. Chromatin was isolated from the rest of the nuclear suspensions. In experiment B, 0.3 pmol of [^{3}H]-dexamethasone-receptor complexes was incubated with mammary gland nuclei or chromatin (0.1 mg DNA) at 0°C for 90 min in a total volume of 0.5 ml.

When expressed as the amount of receptor complex per DNA, about 60% of the form I receptor and more than 65% of the form II were recovered in the chromatin fraction.

In the other type of experiment, chromatin samples were isolated from mammary gland nuclei which had not been incubated with receptors. Subsequently, the receptor complexes were incubated with the chromatin samples, and the amounts of the complexes bound to the chromatin were determined. Table I shows that the two forms have similar binding ability to chromatin: the ability of form I was about 70% of its binding to nuclei, while form II showed the same binding ability to nuclei and chromatin. These data show that most of both receptors bind to chromatin in nuclei.

II. CASEIN mRNA INDUCTION

In vivo studies including ours have demonstrated that the regulation of casein synthesis is dependent on the level of the specific mRNA coding for this protein (*16*, *20*, *23*). An early study using organ culture of the mammary gland argued that glucocorticoid exerts only a potentiation of the essential action of prolactin on intracellular casein mRNA accumulation (*18*). However, it has recently been established that, at least in rodents, glucocorticoid as well as prolactin is essential for casein mRNA induction (*4*, *10*). In addition, nuclear levels of glucocorticoid-receptor complexes are closely correlated with casein mRNA induction (*13*). As shown in the preceding section, most glucocorticoid-receptor complexes in the nuclei of the mammary gland are associated with chromatin and, therefore, chromatin appears to play an important role in casein mRNA induction. In this section, the effects of sodium butyrate, which is known to modulate chromatin structure by inhibiting nuclear histone deacetylase, were investigated in relation to the hormone-mediated casein mRNA induction in cultured rat mammary glands (*9*).

Rat casein has four components (*5*, *6*). We purified mRNA coding for C2-casein (rat 42 K-casein) from lactating rat mammary glands (*16*). C2-casein-specific cDNA was synthesized (*16*) and cloned (*17*). C2-casein mRNA in cultured mammary glands was analyzed by $Rot_{1/2}$ analysis using

TABLE II
Effects of Hormones and Sodium Butyrate on Casein mRNA Levels in the Mammary Gland

Time of initial culture	Conditions of second culture			Casein mRNA (%)
	Hormones	Butyrate	Time	
Experiment 1				
24 hr	—	—	—	0.014
"	I	—	24 hr	0.009
"	I+F	—	"	0.012
"	I+Prl	—	"	0.014
"	I+F+Prl	—	"	0.045
"	I+F+Prl	—	48 hr	0.082
Experiment 2				
48 hr	—	—	—	0.010
"	I+F+Prl	—	24 hr	0.033
"	I+F+Prl	—	48 hr	0.092
Experiment 3				
24 hr	I	—	24 hr	0.010
"	I+F+Prl	—	"	0.088
"	I+F+Prl	2.5 mM	"	0.014
"	I+F+Prl+Pg	—	"	0.048
"	I+F+Prl+Pg	2.5 mM	"	0.010

After the initial culture, mammary glands were cultured under the indicated conditions. The percentage of casein mRNA was determined relative to $Rot_{1/2}$ of purified C2-casein mRNA. I, F, Prl, and Pg represent insulin (5 μg/ml), cortisol (1 μg/ml), prolactin (5 μg/ml), and progesterone (1 μg/ml), respectively.

cDNA or by blot hybridization using a cloned cDNA probe. Mammary glands of 12 day-pregnant rats in their first pregnancy were removed aseptically, cut into small pieces (about 1 mm^3), and cultured in Medium A (Medium 199/Hanks salt/Hepes (pH 7.5)/insulin) in the presence of several combinations of hormones (insulin only, insulin+prolactin, insulin+cortisol, insulin+prolactin+cortisol). RNA was isolated from the cultured glands, electrophoresed on agarose gels, and blotted onto a nitrocellulose filter. C2-casein mRNA was visualized by hybridization with a ^{32}P-labeled cDNA clone and subsequent autoradiography. No difference was seen in the size of casein mRNA among the four different hormone conditions, but the intensity was markedly greater with the conditions including all three hormones than with other conditions (unpublished data). The same was true when the specific RNA was quantified by RNA excess hybridization in solution. As shown in Table II, casein mRNA was fully induced by the coexistence of cortisol and prolactin in the presence of insulin, while mRNA induction was very slight with the single addition of cortisol or prolactin.

The casein mRNA induction by cortisol plus prolactin was strongly inhibited by the presence of sodium butyrate in the culture mediun. Progesterone has previously been shown to inhibit casein mRNA induction (*22*). Whether butyrate blocks the effect of progesterone was also tested, but inhibition effects by the two substances were additive (Table II). Therefore, butyrate appears to block the hormone-mediated induction, but not the repression, of casein mRNA.

The time course of the mRNA induction was examined in Medium A containing cortisol and prolactin, or these hormones plus butyrate. In the first medium, the casein mRNA level increased with incubation time (5-fold during 48 hr incubation). In contrast, in the second medium, no casein mRNA induction was seen during 24 hr incubation. However, the casein mRNA increased 6-fold by incubating the glands for another 48 hr in the butyrate-free medium (Medium A containing cortisol and prolactin). The butyrate-dependent inhibition of casein mRNA induction thus appears reversible.

The effects of butyrate concentration were examined by incubating the glands with different concentrations of the substance under four different hormone conditions. Figure 2 shows that in the medium containing insulin only, the endogenous casein mRNA levels were sensitively decreased by butyrate; for example, the inhibition percentages by 0.5 mM and 3.5 mM butyrate were 40% and 72%, respectively. The inhibition profile in the presence of

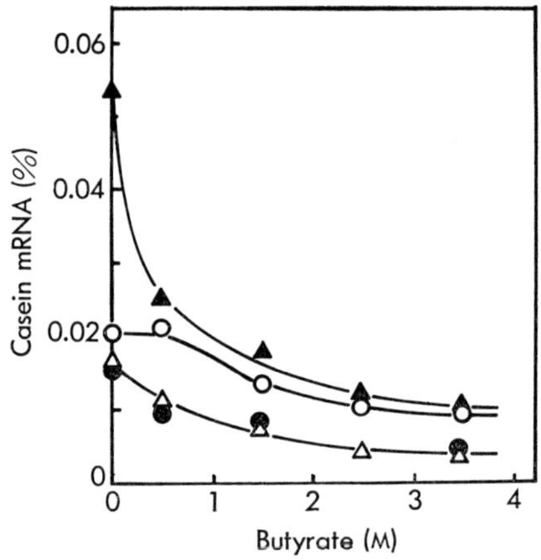

Fig. 2. Dose-dependent inhibition by sodium butyrate. After the initial culture for 24 hr in the presence of insulin, the glands were cultured for 24 hr in the presence of indicated concentrations of sodium butyrate under four different hormone conditions: insulin only (●); insulin and cortisol (△); insulin and prolactin (○); insulin, cortisol, and prolactin (▲). RNA was isolated from the cultured glands, and casein mRNA was quantified.

insulin and cortisol was very similar to that in the presence of insulin alone; therefore, cortisol apparently does not protect the butyrate-dependent inhibition. In contrast, in the presence of insulin and prolactin, casein mRNA levels were less sensitive to butyrate, and little inhibition was seen in the presence of 0.5 mM butyrate. In the medium containing the three hormones, the casein mRNA level was sensitively decreased by the presence of 0.5 mM butyrate. Further, the inhibition percentages approached those in the presence of insulin and prolactin with an increase in butyrate concentration. If there are distinct sites for cortisol and prolactin action with respect to the casein mRNA induction, therefore, butyrate may block the sites for cortisol rather than those for prolactin.

III. HISTONE ACETYLATION IN NUCLEI

The preceding data suggest that modulation in the metabolism of histone acetyl groups has specific importance in casein mRNA induction, since sodium butyrate is known to be a potent inhibitor for this metabolism (*1*, *3*).

Accordingly, in this section, the acetylation of histones was examined in the organ culture system of rat mammary glands (*9*).

Histones were labeled *in vivo* by incubating the glands, which had been cultured in the presence of the three hormones, with radiolabeled acetate for 60 min in the presence or absence of butyrate. Radiolabeled acetate was mainly incorporated into histones H3 and H4, and slightly into histones H2A and H2B. Incorporation into the former two histones was greater in the presence of butyrate than in its absence (H3, 1.4-fold; H4, 2.7-fold). The multiacetylated forms of histone H4 were seen only in the presence of butyrate. Histone deacetylases, but not histone acetylases, have been shown to be inhibited by butyrate during an incubation time as brief as 60 min (*3*); therefore, the turnover time of multiacetylated forms of histone H4 appears to be less than 60 min. When histones labeled with [^{3}H]-acetate in the absence of butyrate were chased for 2 hr, the decrease in intensity of monoacetylated forms of histones H3 and H4 was only 34% and 36%, respectively. The metabolism of acetyl groups in monoacetylated forms of these histones may be slow in the mammary gland.

In experiments to determine whether hormones modulate the metabolism of histone acetyl groups, cell-free conditions were employed for histone acetylation to avoid the problem of hormone effects *in vivo* on intracellular specific activities of the radiolabeled substrate. Nuclei were isolated from the glands, which had been cultured in the presence of insulin only or in the presence of insulin, cortisol, and prolactin, and were then incubated with a known concentration of [^{3}H]-acetyl-CoA in the presence of butyrate. Histones labeled with [^{3}H]-acetate by endogenous acetylases were analyzed by gel electrophoresis. Similar to the acetylation *in vivo*, radiolabeled acetate was incorporated into the nucleosomal core histones under cell-free conditions. The intensity of fluorographs was greater with histones from glands cultured in the three hormones than with those from glands cultured in insulin only (H2A+H2B, 1.4-fold; H3, 1.4-fold; H4, 1.5-fold), while in the intensity of Coomassie-stained histones no significant difference was seen between the two hormonal conditions. Nuclear histone acetylation thus appears greater in mammary glands cultured under the former conditions in which casein mRNA is fully induced.

Acetylation in isolated nuclei was further investigated in relation to casein mRNA induction. Nuclei isolated from glands which had been cultured in different mediums were incubated with [^{3}H]-acetyl-CoA in the presence or

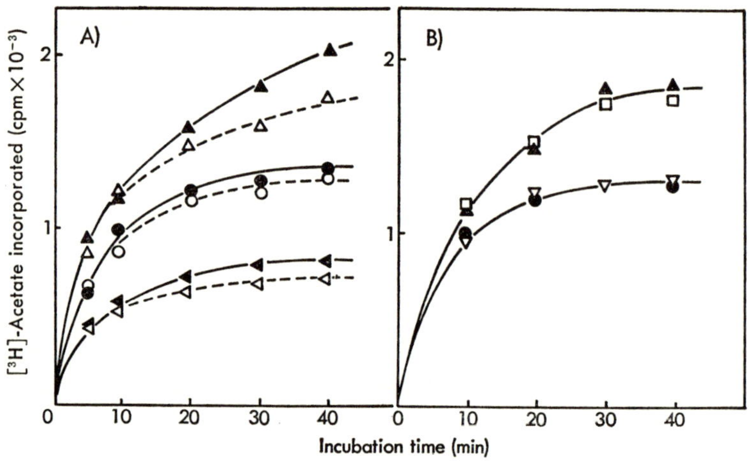

Fig. 3. Effects of hormones and sodium butyrate on histone acetylation. (A) After the initial culture, mammary glands were cultured for 24 hr in the presence of insulin (○, ●); for 24 hr in insulin, cortisol, and prolactin (△, ▲); or for 24 hr in insulin, cortisol, prolactin, and 2.5 mM sodium butyrate and then for 2 hr in a butyrate-free medium containing the three hormones (◁, ◀). Nuclei were isolated from the cultured gland, then *in vitro* acetylation was carried out in the presence (●, ▲, ◀) or absence (○, △, ◁) of 7 mM sodium butyrate. (B) After the initial culture, the glands were cultured for 24 hr in the presence of insulin (●), insulin and prolactin (▽), insulin and cortisol (□), or insulin, cortisol, and prolactin (▲). *In vitro* acetylation was carried out in the same way in the presence of butyrate. Incorporation of radiolabeled acetate into histones was determined by the phosphocellulose technique.

absence of butyrate, the direct inhibitor for deacetylases. The activities incorporating [^{3}H]-acetate in the presence of butyrate (activity A) and in the absence of butyrate (activity B) were determined by phosphocellulose technique. Histone acetylases were estimated from activity A. Histone deacetylase activities were estimated from the percentage of the difference between activities A and B *vs.* activity A. As shown in Fig. 3A, nuclear histone acetylase activities were greater (*e.g.*, about 1.5-fold at 40 min incubation) with the glands cultured in the three hormones than with those cultured in insulin alone. In addition, the histone deacetylase activities were markedly greater in glands cultured under the former conditions (insulin+cortisol+prolactin, 14%; insulin only, 4% at 40 min incubation). Therefore, the metabolism of the histone acetyl groups appears to be stimulated by the coexistence of insulin, cortisol, and prolactin which is required for casein mRNA induction.

The data in the preceding section indicate that hormone-dependent casein mRNA induction is strongly inhibited by butyrate. In a previous report, histone acetylase activities repressed by 20 hr exposure of HTC cells to butyrate were shown to be reversed only gradually following removal of the cells from the inhibitor (*3*); but the direct inhibition of deacetylases can be rapidly reversed upon removal of the cells from butyrate (*3*). Effects of butyrate on histone acetylases were examined by incubating the glands with this fatty acid for 24 hr, then with butyrate-free medium for 2 hr to reverse the direct butyrate inhibition of histone deacetylases. Figure 3A shows that these incubations resulted in a marked decrease in the nuclear acetylation rates (40% of the data in the presence of the three hormones at 40 min incubation). In contrast, the nuclear histone deacetylase activities appeared to be the same as in the presence of the three hormones (14% at 40 min incubation). Therefore, the decreased acetylation rate is probably not due to the increased modification of histones with cold acetate, but to the decreased acetylase activity. It is likely that both the acetylase and the deacetylase activities are repressed during incubation of the glands with butyrate. Taken together with the data on hormonal effects on nuclear acetylation, it can be concluded that changes in casein mRNA levels in the mammary glands are closely correlated with modulation of the metabolism of histone acetyl groups.

Determination of which hormone is responsible for the stimulation of metabolism of histone acetyl groups was made in the following manner. After the glands were cultured in the presence of insulin only, insulin and cortisol, insulin and prolactin, or insulin, cortisol, and prolactin, endogenous histone acetylase activities in isolated nuclei were determined in the presence of butyrate by the process described above. As shown in Fig. 3B, the combination of insulin and cortisol stimulated nuclear histone acetylase activities to the same extent as the combination of the three hormones, while the combination of insulin and prolactin showed no stimulation effect compared with insulin only. Therefore, the steroid hormone appears to have a predominant role in the stimulation of nuclear histone acetylation. The casein mRNA accumulation cannot be fully induced by the combination of insulin and prolactin (Table I). Stimulation of the metabolism of histone acetyl groups may be one of the prerequisite biochemical processes for the induction of casein mRNA.

Although it has been proposed that increased levels of nuclear histone acetylation are correlated with transcriptional activity of chromatin, the lack of correlation has been pointed out in some tissues. A recent model proposed by Perry and Chalkley for the role of histone acetylation includes the importance in transcriptional processes of continuous redistribution of DNA sequences with respect to acetylated chromatin (*21*); such redistribution depends on rapid acetylation and deacetylation. Our present data which show the importance of stimulation of histone acetate metabolic activities in casein mRNA induction appear to be in the same context as the model.

SUMMARY

The nuclear mechanism of hormone action was investigated using cell-free or organ culture systems of the rat mammary gland. Mammary cytoplasmic glucocorticoid-receptor complexes, which had been exposed to a high concentration of ammonium sulfate, were separated into two receptor forms, DNA-cellulose-bound and non-DNA-cellulose-bound receptors. Experiments under cell-free conditions revealed that not only the DNA-cellulose-bound receptors but also the non-DNA-cellulose-bound receptors have binding abilities to nuclei and to nuclear chromatin.

In mammary organ culture systems, casein mRNA was fully induced by the coexistence of glucocorticoid and prolactin. Sodium butyrate, which is known to modulate chromatin structure by inhibiting nuclear histone deacetylases, reversibly and dose-dependently inhibited the hormone-mediated induction of casein mRNA. The inhibition pattern in the presence of several combinations of hormones suggested that butyrate blocks the sites of glucocorticoid, but not of prolactin action.

The metabolic activities of nuclear histone acetyl groups were estimated using nuclei isolated from the mammary gland which had been cultured under different hormone conditions. Nuclear histone acetylases as well as deacetylases were stimulated by the coexistence of prolactin and glucocorticoid or by the presence of glucocorticoid alone. However, the single addition of prolactin to the culture medium showed no stimulation effect.

These results are consistent with a model in which the stimulation of histone acetyl groups is a necessary but insufficient biochemical reaction for casein mRNA induction. Chromatin bindings of glucocorticoid-receptor complexes might be a trigger for this stimulation, and prolactin probably acts on some sites other than histone acetylation. Consequently, the coexistence of glucocorticoid and prolactin is essential for the induction of casein mRNA.

REFERENCES

1) Boffa, L.C., Vidali, G., Mann, R.S., and Alfrey, V.G. (1978) Suppression of histone deacetylation *in vivo* and *in vitro* by sodium butyrate. *J. Biol. Chem.* **253**, 3364–3366.
2) Catt, K.J. and Dufau, M.L. (1977) Peptide hormone receptors. *Annu. Rev. Physiol.* **39**, 529–557.
3) Covault, J., Sealy, L., Schnell, R., Shires, A., and Chalkley, R. (1982) Histone hypoacetylation following release of HTC cells from butyrate. *J. Biol. Chem.* **257**, 5809–5815.
4) Ganguly, R., Ganguly, N., Mehta, N.M., and Banerjee, M.R. (1980) Absolute requirement of glucocorticoid for expression of the casein gene in the presence of prolactin. *Proc. Natl. Acad. Sci. U.S.A.* **77**, 6003–6006.
5) Hirose, M., Kato, T., Omori, K., Maki, M., Yoshikawa, M., Sasaki, R., and Chiba, H. (1981) Purification and properties of a major casein component of rat milk. *Biochim. Biophys. Acta* **667**, 309–320.
6) Hirose, M., Kato, T., Omori, K., Takeuchi, M., Yoshikawa, M., Sasaki, R., and Chiba, H. (1981) Purification and characterization of four components of rat caseins. *Biochim. Biophys. Acta* **671**, 139–145.
7) Hirose, M., Maki, M., and Chiba, H. (1980) Characterization of a progestin-binding component in lactating rat mammary glands. *Biochim. Biophys Acta* **629**, 168–177.
8) Hirose, M., Ohshima, A., Maki, M., and Chiba, H. (1983) Nuclear binding and sedimentation properties of two activated forms of glucocorticoid-receptor complexes from rat mammary glands. *Biochim. Biophys. Acta* **760**, 84–91.
9) Hirose, M., Sarui, K., and Sunagawa, A. (1985) Correlation between nuclear histone acetylation and casein messenger RNA induction in the mammary gland. *J. Biochem.* **97**, 781-789.
10) Kulski, J.K., Topper, Y.J., Chomczynski, P., and Qasba, P. (1983) An essential role for glucocorticoid in casein gene expression in rat mammary explants. *Biochem. Biophys. Res. Commun.*, **114**, 380–387.
11) Leach, K.L., Dahmer, M.E., Hammond, N.D., Sando, J.J., and Pratt, W.E. (1979) Molybdate inhibition of glucocorticoid receptor inactivation and transformation. *J. Biol. Chem.* **254**, 11884–11890.
12) Le Fevre, B., Bailly, A., Sallas, N., and Milgrom, E. (1979) Activated steroid-

receptor complex. Comparison of assays using DNA-cellulose or homologous nuclei. *Biochim. Biophys. Acta* **586**, 266–272.
13) Majumder, P.K., Joshi, J.B., and Banerjee, M.R. (1983) Correlation between nuclear glucocorticoid receptor levels and casein gene expression in murine mammary gland *in vitro*. *J. Biol. Chem.* **258**, 6793–6798.
14) Maki, M., Hirose, M., and Chiba, H. (1980) Alterations in glucocorticoid receptor conformation by molybdate. *J. Biochem.* **87**, 1851–1854.
15) Maki, M., Hirose, M., and Chiba, H. (1980) Occurrence of common binding sites for progestin and glucocorticoid in the lactating mammary gland of the rat. *J. Biochem.* **88**, 1845–1854.
16) Maki, M., Hirose, M., and Chiba, H. (1981) Effects of weaning and resuckling on casein messenger RNA levels in the rat mammary gland. *Agric. Biol. Chem.* **45**, 2921–2923.
17) Maki, M., Ueda, M., Hirose, M., and Chiba, H. (1982) Synthesis of rat casein antigen in *E. coli*. *Agric. Biol. Chem.* **46**, 1979–1981.
18) Matsuik, R.J. and Rosen, J.M. (1978) Prolactin induction of casein mRNA in organ culture. *J. Biol. Chem.* **253**, 2343–2347.
19) McBlain, W.A., Toft, D.O., and Shyamala, G. (1981) Transformation of mammary cytoplasmic glucocorticoid receptor under cell-free conditions. *Biochemistry* **20**, 6790–6798.
20) Nakhasi, H.L. and Qasba, P.K. (1979) Quantitation of milk proteins and their mRNAs in rat mammary gland at various stages of gestation and lactation. *J. Biol. Chem.* **254**, 6016–6025.
21) Perry, M. and Chalkley, R. (1982) Histone acetylation increases the solubility of chromatin and occurs sequentially over most of the chromatin. A novel model for the biological role of histone acetylation. *J. Biol. Chem.* **257**, 7336–7347.
22) Rosen, J.M., O'Neal, D.L., McHugh, J.E., and Comstock, J.P. (1978) Progesterone-mediated inhibition of casein mRNA and polysomal casein synthesis in the rat mammary gland during pregnancy. *Biochemistry* **17**, 290–297.
23) Rosen, J M., Woo, S.L.C., and Comstock, J.P. (1975) Regulation of casein messenger RNA during the development of the rat mammary gland. *Biochemistry* **14**, 2895–2903.
24) Schmidt, T.J. and Litwack, G. (1982) Activation of the glucocorticoid-receptor complex. *Physiol. Rev.* **62**, 1131–1192.
25) Shyamala, G. (1973) Specific cytoplasmic glucocorticoid receptors in lactating mammary glands. *Biochemistry* **12**, 3085–3090.
26) Smith, P. and Holt, C. (1981) Interaction of the activated cytoplasmic glucocorticoid hormone receptor complex with the nuclear envelope. *Biochemistry* **20**, 2900–2908.

27) Thrall, C.L., Webster, R.A., and Spelsberg, T.C. (1978) Steroid receptor interaction with chromatin. In "The Cell Nucleus," Vol. VI, ed. Busch, H., pp. 461–529, Academic Press, New York.
28) Toft, D. and Nishigori, H. (1979) Stabilization of the avian progesterone receptor by inhibitors. *J. Steroid Biochem.* **11**, 413–416.

7

Glucocorticoid Hormone: Mechanisms of Action, ed. Sakamoto, Y. and Isohashi, F., pp. 157–172,
Japan Sci. Soc. Press, Tokyo/Springer-Verlag, Berlin (1986)

ANALYSIS OF THE MOLECULAR ORGANIZATION OF STEROID HORMONE RECEPTOR SYSTEM

AKIRA MURAYAMA

Department of Physiological Chemistry, The Tokyo Metropolitan Institute of Medical Science, Tokyo 113, Japan

The currently accepted dogma of steroid hormone action may be summarized as follows. "Steroid hormones, which are synthesized in the endocrine glands and secreted into the circulation, penetrate the plasma membrane by simple diffusion due to the strong hydrophobicity. Steroid hormone receptors bind specifically with the hormones in the cytoplasm of the target cells, then translocate into the nuclei and bind to the chromatin, and finally regulate the expression of specific genes." Study of the action mechanism of steroid hormones in the target cells is aimed at solving the important problem of the little-known mechanism of regulation of gene expression in the highly organized animal cells. Detailed knowledge of the molecular properties of steroid hormone receptors is basic to the study of the hormone action mechanism. More than twenty years of intensive investigative effort on the biochemistry of these receptors has led to many interesting proposals on their molecular constitution (*1–3*, *8*, *9*, *11*, *12*, *24*, *25*, *29*). However, there remain many discrepancies between the various proposed molecular receptor models, and little is known of their reaction in the target cells which induces the hormonal response.

Two major obstacles seem to have been disregarded in the study of steroid hormone receptors (*12*). First, these receptors undergo complicated association and dissociation with the coextracted protein factors, an appropriate understanding of which is an important basis of the receptor phenomena (*1, 2, 4, 9, 12, 22, 23*). Second, steroid hormone receptors are extremely susceptible to modifications by the proteases coextracted with them (*4, 7, 18, 20, 22*). Through the slight proteolytic modification, the binding site of the receptors to the hormones is unaffected, while other unique molecular properties of the receptors, which are thought to be responsible for the biological functions, are lost (*4, 7, 18, 20, 22*). The binding site to the hormone is only a part of the receptor molecule. It is important to protect the steroid receptors from the proteases, and to separate them from the protein factors which associate with them in great numbers.

We recently presented new information on the molecular constitution of the estrogen receptor (ER)* system (*4, 18, 19, 21, 23*) and progesterone receptor system (*22*) showing their similarity. We showed that there is one basic ER molecule (*vero*-ER) (sedimentation coefficient, 4.5S; Stokes radius, 44 Å) in the estrogen target tissues. *Vero*-ER interacted specifically with the endogenous protein factors designated as ER-binding factors (ERBFs) ("5S" ER-forming factor ("5S" ER-FF), (component A); "6S" ER-FF, $(\text{component B})_6$; "8S" ER-FF, (component A)•$(\text{component B})_6$) to form ERs ("5S" ER, (*vero*-ER)•("5S" ER-FF); "6S" ER, (*vero*-ER)•("6S" ER-FF); "8S" ER, (*vero*-ER)•("8S" ER-FF)) with various molecular constitutions. *Vero*-ER is very labile to the coextracted proteases and yields receptor fragments (*secto*-ER (sedimentation coefficient, 4.5S; Stokes radius, 35 Å); "4.2S" ER; "3.8S" ER, "2.9S" ER, *etc.*), which still bind with estradiol but no longer bind with ERBFs (Fig. 1).

This article reviews the recent results of our laboratory on the study of the molecular organization of the ER system, and summarizes current infor-

---

*The abbreviations used are: ER, estrogen receptor; ERBF, estrogen receptor-binding factor; "8S" ER-FF, "8S" ER-forming factor; "6S" ER-FF, "6S" ER-forming factor; "5S" ER-FF, "5S" ER-forming factor; [^{3}H]-estradiol, [2,4,6,7,16,17-^{3}H]-estradiol; TEM buffer, 10 mM Tris-HCl, 1.5 mM EDTA, 1.5 mM 2-mercaptoethanol, pH 8.0; TEMA buffer, TEM buffer containing 0.25 mM antipain (*28*); PEMA buffer, 30 mM potassium phosphate, 1.5 mM EDTA, 1.5 mM 2-mercaptoethanol, 0.25 mM antipain, pH 6.5.

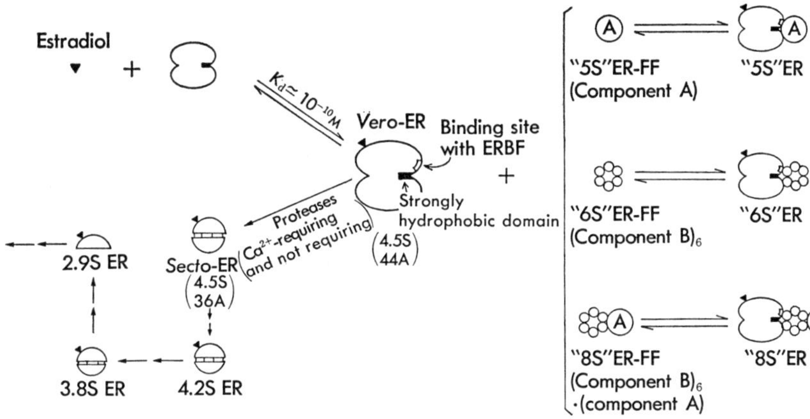

Fig. 1. Schematic representation of the molecular constitutions of various forms of ERs and ERBFs.

mation on the unique molecular properties of the receptor (*5*, *13–16*, *20*). It appears highly possible that the fundamental features of the ER system described in this article may also be found, with some modifications, in the receptor systems of other steroid hormones including glucocorticoids (*7*, *22*).

I. THE CYTOSOLIC ER SYSTEM (*4*, *18*, *19*, *21*, *23*)

The process of elucidation of the molecular constitution of various forms of ERs and ERBFs as schematically represented in Fig. 1 will be briefly summarized in this section.

The ER of fresh uterine (cow or gilt) cytosol was sedimented at 8S ("8S" ER) under low salt conditions, and at 4.5S in the presence of 0.4 M KCl or 0.4 M NaSCN. Similar remarkable variations have been reported of the sedimentation pattern of steroid hormone receptors depending on salt conditions, and a molecular model could explain this variation. The basic constituents of the receptor system should be separated and the association and dissociation reconstituted. In an analysis of this system, the receptor must be protected from the coextracted proteases, since, as stated, with even the slightest proteolytic modification, the receptor loses its unique property to vary the sedimentation pattern depending on the salt conditions. We effec-

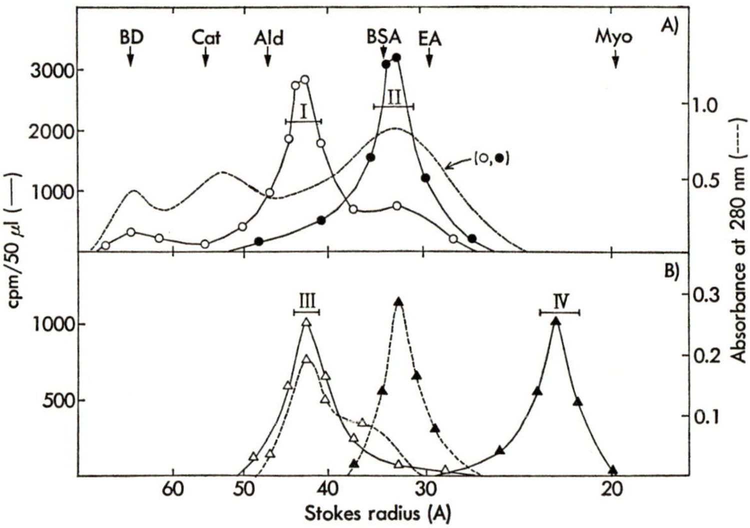

Fig. 2. Analysis of *vero*-ER and *secto*-ER by gel filtration. (A) Labeled porcine uterine cytosol was subjected to gel filtration on a Sephadex G-150 column (2.6 ×96 cm) in TEMA buffer containing 0.4 M KCl (○), or in TEM buffer containing 0.4 M KCl (●). (B) Gel filtration of fraction I (*vero*-ER) (△) or fraction II (*secto*-ER) (▲) was carried out on a Sephadex G-150 column (2.6×96 cm) in TEMA buffer containing 0.4 M NaSCN. ——, [^{3}H]-estradiol content; ----, absorbance at 280 nm. The arrows indicate the peaks of elution for Blue Dextran (BD), catalase (Cat), aldolase (Ald), bovine serum albumin (BSA), egg albumin (EA), and myoglobin (Myo).

tively protected the uterine cytosolic ER from the endogenous proteases by antipain (28), a protease inhibitor of microbial origin.

When the labeled cytosol prepared in TEMA (10 mM Tris-HCl, .15 mM EDTA, 1.5 mM 2-mercaptoethanol, 0.25 mM antipain, pH 8.0) buffer was subjected to a two-step gel filtration on a Sephadex G-150 column in TEMA buffer in the presence of 0.4 M KCl, and then in the presence of 0.4 M NaSCN, ER with a Stokes radius of 44 Å (fraction III, Fig. 2B) and a sedimentation coefficient of 4.5S was obtained as the sole ER activity in the eluate. This indicated strongly that there is one basic ER molecule in the uterine cytosol. The basic receptor molecule with a Stokes radius of 44 Å and a sedimentation coefficient of 4.5S was designated as *vero*-ER (*vero*, taken from the Latin *verus* for true).

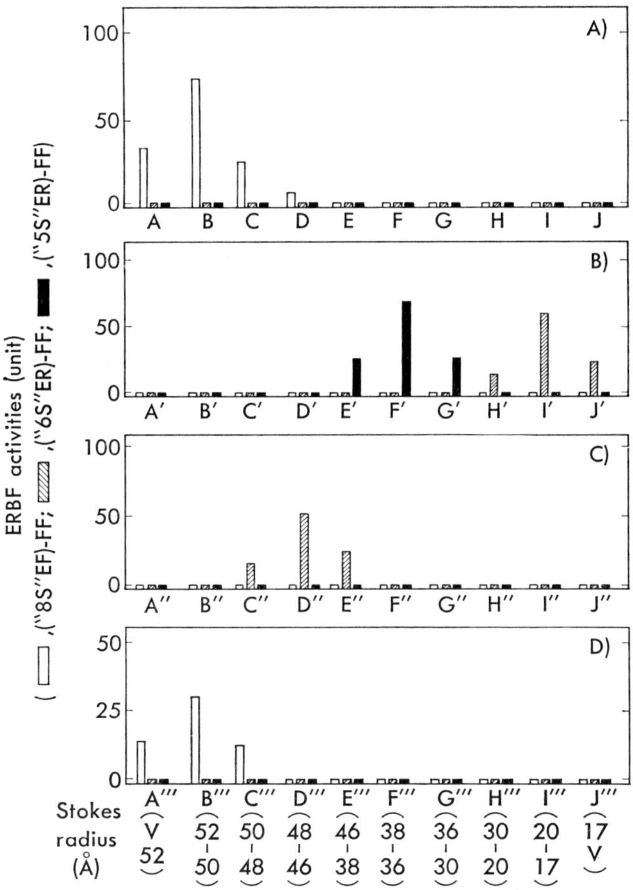

Fig. 3. Analysis of ERBFs by gel filtration. Gel filtration was carried out on a Sephadex G-150 column (1.6×96 cm). The eluate was divided into fractions, desalinated, concentrated, and then assayed for ERBF activity (*23*). (A) Gel filtration of the unlabeled cytosol in TEM buffer. (B) Gel filtration of "8S" ER-FF (fraction B in Fig. 3A) in TEM buffer containing 0.4 M NaSCN. (C) Gel filtration of component B (fraction I′ in Fig. 3B) in TEM buffer. (D) Gel filtration of mixture of "5S" ER-FF (fraction F′ in Fig. 3B) and "6S" ER-FF (fraction D″ in Fig. 3C) in TEM buffer.

Vero-ER by itself did not undergo self-association and -dissociation depending on the salt conditions, but when *vero*-ER was mixed with unlabeled cytosol, ER sedimented at 8S under low salt conditions, suggesting the presence of a factor ("8S" ER-FF) which binds with *vero*-ER to form "8S" ER. When the unlabeled cytosol prepared in TEM buffer was subjected to gel filtration on a Sephadex G-150 column under low salt (TEM) conditions, "8S" ER-FF was eluted in the fractions with a Stokes radius of around 51 Å (fraction B, Fig. 3A). "8S" ER-FF is a protein, which does not bind with estradiol; it was shown to be composed of a subunit structure, (component A)•(component $B)_6$, in the following manner. "8S" ER-FF was dissociated into component A (Stokes radius, 37 Å) (fraction F′, Fig. 3B), and component B (Stokes radius, 18.5 Å) (fraction I′, Fig. 3B) by gel filtration in the presence of 0.4 M NaSCN. Component A bound with *vero*-ER to form ER sedimenting at 5.5S ("5S" ER) in TEMA buffer with or without 0.4 M KCl. Component A was further designated as the "5S" ER-FF. Component B underwent self-association under low salt (TEM) conditions to form the hexamer (Stokes radius, 47 Å) (fraction D″, Fig. 3C). The hexamer of component B bound with *vero*-ER to form "6S" ER (sedimentation coefficient, 6.5S) under low salt (TEMA) conditions, and was designated as the "6S" ER-FF. When "5S" ER-FF and "6S" ER-FF were mixed and then subjected to gel filtration under low salt (TEM) conditions, "8S" ER-FF merged with a concomitant disappearance of "5S" ER-FF and "6S" ER-FF (Fig. 3D). "5S" ER-FF, "6S" ER-FF, and "8S" ER-FF were designated as ERBFs.

Reconstructed "6S" ER ((*vero*-ER)•(component $B)_6$) bound with "5S" ER-FF (component A) to form "8S" ER; however, reconstructed "5S" ER ((*vero*-ER)•(component A)) did not bind with "6S" ER-FF ((component $B)_6$). Accordingly, it was assumed that in "8S" ER, component A is bound to *vero*-ER not directly, but *via* component B. Since the protein factors which bind specifically with *vero*-ER are present in large molar excess to *vero*-ER in the normal target tissues (*4*), it appears more appropriate to designate them as receptor-binding factors than as receptor subunits.

When gel filtration of the labeled cytosol was carried out in TEM (TEMA devoid of antipain) buffer in the presence of 0.4 M KCl, ER eluted in the fractions (fraction II, Fig. 2A) with a smaller Stokes radius (35 Å) as compared to *vero*-ER. This indicated that *vero*-ER was proteolyzed by a protease during the gel filtration in the absence of antipain. The proteolyzed ER sedimented at 4.5S under low salt (TEMA) conditions similar to *vero*-ER, but no longer possessed the capability to bind with ERBFs. The proteolyzed ER with a Stokes radius of 35 Å and a sedimentation coefficient of 4.5S was designated as *secto*-ER (*secto*, taken from Latin *secare* for cut). *Secto*-ER

TABLE I
Molecular Parameters of ERs and ERBFs

	Symbolic sign	Sedimentation coefficient (S)	Stokes radius (A)	Molecular weight[a]	Frictional ratio[a] (f/f_0)
Vero-ER	(R)	4.5	44	82,000	1.53
"5S" ER-FF (component A)	(A)	3.8	37	58,000	1.45
Component B	(B)	1.8	18.5	13,700	1.17
"6S" ER-FF	(B_6)	4.1	47	80,000	1.65
"8S" ER-FF	(B_6)(A)	6.9	51	145,000	1.47
"5S" ER	(R)(A)	5.5	60	136,000	1.77
"6S" ER	(R)(B_6)	6.5	60	162,000	1.67
"8S" ER	(R)(B_6)(A)	8.0	68	225,000	1.69
Secto-ER		4.5	35	65,000	1.32
"2.9S" ER		2.9	24	29,000	1.18

[a] Calculated from the values of sedimentation coefficient and Stokes radius after Siegel and Monty (*26*) assuming partial specific volume to be 0.725 cm^3/g after Martin and Ames (*10*).

sedimented at 2.9S in the presence of 0.4 M NaSCN. When *secto*-ER was subjected to gel filtration in the presence of 0.4 M NaSCN, ER was eluted in the fractions with a Stokes radius of 24 Å (fraction IV, Fig. 2B). The ER of fraction IV sedimented at 2.9S in the presence of NaSCN, and at 4.5S in the absence of NaSCN. These results indicated that in the presence of 0.4 M NaSCN, *secto*-ER is dissociated at the nick to yield the "2.9S" ER fragment (Stokes radius, 24 Å) and a counterpart of similar molecular size.

Further proteolytic modifications of *vero*-ER by the coextracted enzymes yielded various ER fragments ("4.2S" ER, "3.8S" ER, *etc.*) depending on the conditions. Proteolysis of *vero*-ER was effected by two different types of cytosolic proteases, namely, Ca^{2+}-requiring and -not requiring proteases (*20*); antipain protected *vero*-ER effectively from both these proteases (*20*).

The molecular parameters and molecular constitutions of the various forms of estrogen receptors and ER-binding factors are summarized in Table I.

II. MOLECULAR MECHANISM OF THE ACTIVATION OF ER (*13–15*)

Steroid hormone receptors are assumed to bind with the hormones in the cytoplasm of the target cells, and then undergo the activation process to acquire the capability to translocate from the cytoplasm into the nuclei (*8*, *27*). Studies on the activation process of the uterine ER strongly indicated that *vero*-ER is the activated ER which translocates from the cytoplasm into the nucleus and interacts with chromatin (*15*).

When uterine slices were incubated with [^{3}H]-estradiol at near 0°C, and then analyzed for the amounts of ER present in the cytosol and nuclei, all of the receptor bound with [^{3}H]-estradiol was present in the cytosol fraction (*8*, *13*, *27*). However, when incubation of the uterine slices was carried out at an elevated temperature (37°C), the amount of cytosolic ER decreased with a concomitant increase in the amount of nuclear receptor (*8*, *13*, *27*). These results strongly indicated the presence of a temperature-dependent activation process of the receptor (*8*).

About 60 % of the nuclear ER bound with [^{3}H]-estradiol was extractable with TEM buffer containing 0.4 M KCl. Molecular parameters of the extracted ER coincided with those of *vero*-ER when analyzed in the presence of 0.4 M NaSCN by destroying the interaction with ERBFs (*13*). This was strong indication that *vero*-ER translocates into the nuclei as the intact form without undergoing proteolytic modification by the cytoplasmic proteases.

On the other hand, a protease which attacks *vero*-ER was extracted from the isolated nuclei with TEM buffer containing 0.4 M NaSCN (*13*). *Vero*-ER could not be protected from the nuclear protease with antipain. The biological significance of this nuclear protease in relation to the receptor function in chromatin remains to be elucidated. All of the nuclear ER bound with [^{3}H]-estradiol was extractable with TEM buffer containing 0.4 M NaSCN, but the extracted ER had sustained proteolytic modification by the coextracted enzyme (*13*).

The basic subunits of ERBFs, components A and B, were also extracted from the nuclei (*13*). Those extracted with TEM buffer containing 0.4 M NaSCN amounted to about 25 % of the level in the cytosol (*13*).

We then estimated the molecular form of the activated ER by analyzing the capability of various forms to translocate into the isolated nuclei (*13*). To eliminate the artificial complications, we used a minimum amount of nuclei *vs.* ERs, and incubated them carefully at 2°C for 3 hr in Tris buffer (pH 8.0) in the presence of 0.32 M sucrose and 1 mM $MgCl_2$ to protect the nuclei (*30*). The amount of ER per nucleus under the incubation conditions was the same as the average ER content per cell of the porcine uterus. The amount of ER translocated into the nuclei was estimated by subtracting the amount

TABLE II
Comparison of the Accumulation of Various Forms of ERs in the Isolated Porcine Nuclei

	Receptor bound to the nuclei[a] (fmol)	Receptor adsorbed to the nuclear envelopes[a] (fmol)	Receptor translocated into the nuclei[b] (fmol)	Translocated (%)
Vero-ER•[^{3}H]-E	226	35	191	75
"5S" ER•[^{3}H]-E	65	45	20	8
"6S" ER•[^{3}H]-E	110	37	73	29
"8S" ER•[^{3}H]-E	100	40	60	24
("5S" ER•[^{3}H]-E) •(component C)	133	110	23	9
Secto-ER•[^{3}H]-E	85	68	17	7
"3.8S" ER•[^{3}H]-E	74	55	19	8

[a] Estrogen receptor (250 fmol) labeled with [^{3}H]-estradiol was incubated with the nuclei or nuclear envelopes in 0.32 M sucrose-1 mM $MgCl_2$-10 mM Tris-HCl-0.25 mM antipain, pH 8.0, for 3 hr at 4°C. [b] Obtained by subtracting the amount of receptor adsorbed to the nuclear envelopes from that of receptor bound to the nuclei.

of ER bound to the nuclear envelopes from that of ER bound to the whole nuclei. As summarized in Table II, *vero*-ER•E (*vero*-ER bound with estradiol) possessed a particularly high capability to translocate into the nuclei as compared to the complexes of *vero*-ER with ERBFs ("5S" ER•E, "6S" ER•E and "8S" ER•E), and partially proteolyzed ERs (*secto*-ER•E and "3.8S" ER•E). These results indicated strongly that *vero*-ER•E is the activated ER which translocates into the nucleus, and the translocation is inhibited reversibly through binding with ERBFs (inhibitors), and irreversibly through the proteolytic modifications. It was previously proposed that under the activation conditions of uterine ER, "4S" ER•E (unactivated receptor) is transformed into "5S" ER•E (activated receptor) (*8*). However, translocation of "5S" ER•E into the nuclei was almost nil. It was further found that "5S" ER in the cytosol is joined with a macromolecular component (component C) under the assay conditions of the translocation of ER into the isolated nuclei (*13*). The complex, ("5S" ER•E)•(component C), possessed relatively high affinity for the nuclear envelopes, but did not translocate into the nuclei (Table II). The slight nuclear translocation of ER observed when "6S" ER•E or "8S" ER•E was assayed is due to the partial dissociation of *vero*-ER•E effected by the Mg^{2+} (1 mM) present in the incubation medium. This remarkable effect of Mg^{2+} on the dissociation of *vero*-ER from ERBFs will be described in the following section.

The results described in this section strongly indicated that *vero*-ER•E is the activated ER, in which form ER is translocated from the cytoplasm into the nucleus. *Vero*-ER has an extremely strong hydrophobic domain, which is concealed through binding with ERBFs (see the next section). It is assumed that the strong hydrophobic domain plays an essential role in the process of accumulation in the nuclei. It appears that both *vero*-ER•E and proteolyzed ERs (*secto*-ER•E and "3.8S" ER•E) may pass through the nuclear pores without suffering sterical hindrances. However, the proteolyzed ERs which no longer possess the strongly hydrophobic domain (see the next section) cannot accumulate in the nuclei. The complexes of *vero*-ER with ERBFs ("5S" ER•E, "6S" ER•E, and "8S" ER•E) were assumed to be hindered in passing through the nuclear pores because of their bulky size.

III. SINGULAR PROPERTIES OF THE BASIC ER MOLECULE (*vero*-ER)

*1. Extremely Strong Hydrophobicity of the Basic ER Molecule (*vero-*ER) (16)*

Vero-ER was shown to possess an extremely strong hydrophobic domain, which is concealed through binding with ERBFs as follows. *Vero*-ER dissociated from ERBFs was bound to a phenyl- or octyl-Sepharose column under low salt (TEMA) conditions. The attached *vero*-ER could not be eluted with TEMA buffer containing 50 % ethylene glycol, with which most of the hydrophobic proteins of the cytosol were eluted from the column. To elute *vero*-ER from the column, it was necessary to utilize ethanol as a polarity-reducing agent; it could be eluted quantitatively with TEMA buffer containing 30 % ethanol. Dissociation of [^{3}H]-estradiol from the receptor during the process of the elution from phenyl-Sepharose was negligible.

In contrast, reconstructed "5S" ER, "6S" ER, and "8S" ER passed straight through the phenyl-Sepharose column in TEMA buffer. This indicated that the strongly hydrophobic domain of *vero*-ER is concealed in the complexes with ERBFs. The association and dissociation of *vero*-ER with ERBFs is regulated mainly by ionic forces (see the previous section). Accordingly, it is assumed that the binding sites of *vero*-ER with ERBFs, and the strongly hydrophobic domain occupy different positions on the *vero*-ER molecule.

Both *secto*-ER and "3.8S" ER passed straight through the phenyl-Sepharose column under low salt (TEMA) conditions, indicating that the strong hydrophobic domain of *vero*-ER was destroyed by the proteolytic modifications. The simultaneous removal of the binding sites with ERBFs, and the strong hydrophobic domain suggested that they occupy positions close to each other on the *vero*-ER molecule.

*2. Specific Reactivity of Basic ER Molecule (*Vero*-ER) with ERBFs is Regulated by Multiple Inorganic Ions (5, 6, 14, 17)*

The specific association and dissociation of *vero*-ER with ERBFs were found to be regulated in a complicated way by various mineral ions at lower concentrations. Order of the affinity of *vero*-ER with ERBFs, estimated as apparent dissociation constant (K_d) values under various conditions, is summarized in Table III. In Tris (TEMA) buffer, affinity of *vero*-ER for "5S" ER-FF (component A) was one order higher than the affinity for "8S" ER-FF ((component A)•(component $B)_6$) (*14*). This relatively large difference in the affinities of ERBFs for *vero*-ER reasonably explains the higher stability of "5S" ER as compared to "8S" ER observed in Tris buffer. It has been widely reported that, in Tris buffer containing 0.4 M KCl, "5S" ER remains stable, while "8S" ER is dissociated to yield *vero*-ER ("4S" ER) (*8*, *11–13*, *25*). Irreversible formation of "5S" ER from "8S" ER by treating the latter under certain conditions has also been reported (*8*, *14*, *23*). The following information will show that these phenomena occur only in Tris buffer.

The selection of buffer was found to be very important in the biochemical study of steroid hormone receptors. In phosphate (PEMA) (30 mM potassium phosphate, 1.5 mM EDTA, 1.5 mM 2-mercaptoethanol, 0.25 mM antipain, pH 6.5) buffer, or in TEMA buffer containing phosphate (30 mM), affinity of *vero*-ER for "5S" ER-FF decreased nearly one order, while the affinity for "8S" ER-FF increased over four times as compared to the observations in Tris (TEMA) buffer (*17*). As a result, "8S" ER is more stable than "5S" ER in the presence of phosphate (30 mM) (Table III). In PEMA buffer or in TEMA buffer containing 30 mM phosphate, both "8S" ER and "5S" ER were dissociated into *vero*-ER ("4S" ER) and the corresponding ERBFs when 0.4 M KCl was added (*17*).

It was further found that the specific interaction of *vero*-ER with ERBFs was also regulated by molybdate (*6*). Molybdate, similar to phosphate, increased the affinity of *vero*-ER for "8S" ER-FF, and decreased the affinity for "5S" ER-FF (*6*). In the presence of 30 mM phosphate, the affinity of *vero*-ER for "8S" ER-FF was stronger than that for "5S" ER-FF (Table

TABLE III
Apparent Equilibrium Dissociation Constant of *Vero*-ER with ERBFs

Buffer conditions	K_d (M)		
	"8S"ER-FF	"6S"ER-FF	"5S"ER-FF
TEMA	2.7×10^{-9}	3.5×10^{-9}	2.8×10^{-10}
TEMA + PO_4^{3-} (30 mM) or PEMA	7.5×10^{-10}	8.7×10^{-10}	1.3×10^{-9}
TEMA + MoO_4^{2-} (30 mM)	6.7×10^{-10}	5.9×10^{-9}	3.7×10^{-9}
TEMA + PO_4^{3-} (30 mM) + MoO_4^{2-} (30 mM) or PEMA + MoO_4^{2-} (30 mM)	1.8×10^{-10}	—	—
TMA[a] + Mg^{2+} (1 mM)	3.5×10^{-8}	6.0×10^{-8}	2.8×10^{-9}

[a] TEMA devoid of EDTA.

III). The effects of molybdate and phosphate on the association and dissociation of *vero*-ER with ERBFs were synergetic (*17*). In the coexistence of phosphate (30 mM) and molybdate (30 mM), the affinity of *vero*-ER for "8S" ER-FF was so increased (Table III) that *vero*-ER could not be dissociated from "8S" ER-FF by the addition of 0.4 M KCl (*17*).

Vero-ER dissociated from ERBFs undergoes further complicated interactions with other coextracted cellular components, which greatly disturbs its purification and characterization. Sufficient purification of the receptor in "8S" form and separation of *vero*-ER from ERBFs can avoid these complications. The information summarized in this section includes important criteria for the selection of appropriate buffer conditions in each step of the receptor purification.

It was further found that the specific interaction of *vero*-ER with ERBFs is regulated by Mg^{2+} at physiological concentrations (*5*). The apparent K_d

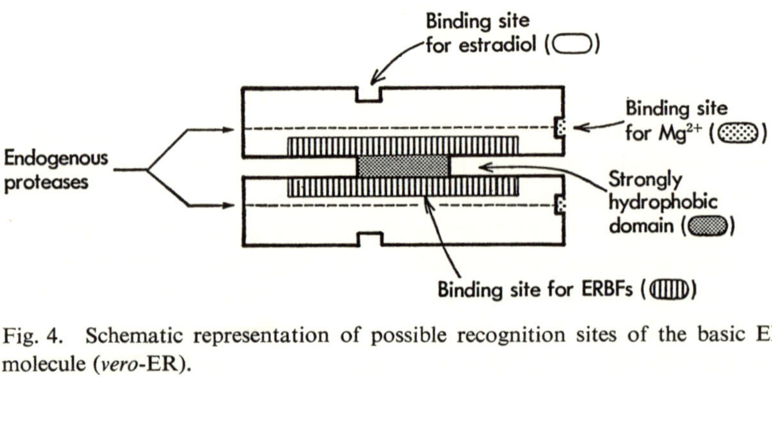

Fig. 4. Schematic representation of possible recognition sites of the basic ER molecule (*vero*-ER).

values of *vero*-ER for "5S" ER-FF, "6S" ER-FF, and "8S" ER-FF in the presence of 1 mM Mg^{2+} were all increased over 10 times that in the absence of Mg^{2+} (Table III). These results suggest that *vero*-ER belongs to the Mg^{2+}-binding proteins. It is expected that Mg^{2+} plays an important role in the action mechanism of steroid hormones.

Important reaction sites of *vero*-ER hitherto identified are schematically represented in Fig. 4.

SUMMARY

One basic ER molecule (*vero*-ER) (Stokes radius, 44 Å; sedimentation coefficient, 4.5S) and protein components designated as ERBFs ("5S" ER-FF, (component A, Stokes radius, 37 Å); "6S" ER-FF (component B, Stokes radius, 18.5 Å)$_6$; "8S" ER-FF, (component A)•(component B)$_6$ are present in the estrogen target tissues which interact specifically with each other to form ERs ("5S" ER, (*vero*-ER)•("5S" ER-FF); "6S" ER, (*vero*-ER)•("6S" ER-FF); and "8S" ER, (*vero*-ER)•("8S" ER-FF)) with various molecular constitutions. *Vero*-ER bound with estradiol was shown to be the activated receptor which is translocated from the cytoplasm into the nucleus, and the translocation was inhibited reversibly through binding with ERBFs (inhibitors). *Vero*-ER was shown to possess an extremely strong hydrophobic domain, which is concealed through binding with ERBFs. The specific association and dissociation of *vero*-ER with ERBFs were regulated by various inorganic ions at lower concentrations. In Tris buffer, affinity of *vero*-ER for "5S" ER-FF was one order higher than the affinity for "8S" ER-FF. Phosphate and molybdate increased the affinity of *vero*-ER for "8S" ER-FF, and decreased its affinity for "5S" ER-FF. In the presence of phosphate or molybdate, "8S" ER was more stable than "5S" ER. The effects of phosphate and molybdate on the stabilization of "8S" ER were synergetic. Mg^{2+} at physiological concentrations decreased remarkably the affinity of *vero*-ER for ERBFs. *Vero*-ER was assumed to be a Mg^{2+}-binding protein. It was extremely susceptible to modifications by endogenous proteases and, with the slightest modification, lost its binding site with ERBFs, its strong hydrophobic domain, and the capability to translocate into the isolated nuclei.

Acknowledgments

I thank cordially Prof. Yukiya Sakamoto and Dr. Fumihide Isohashi for inviting me to this important symposium. This review is dedicated to the memory of the late Prof. Tadashi Yamamoto who inspired and supported the author with his outstanding personality.

REFERENCES

1) Colvard, D.S. and Wilson, E.M. (1981) Androgen receptor-promoting factor that converts the 4.5S form of the androgen receptor to 8S. *Endocrinology* **109**, 496–504.
2) Colvard, D.S. and Wilson, E.M. (1984) Androgen receptor binding to nuclear matrix *in vitro* and its inhibition by 8S androgen receptor promoting factor. *Biochemistry* **23**, 3479–3486.
3) Coty, W.A., Schrader, W.T., and O'Malley, B.W. (1979) Purification and characterization of the chick oviduct progesterone receptor A subunit. *J. Steroid Biochem.* **10**, 1–12.
4) Fukai, F. and Murayama, A. (1981) Cytoplasmic estrogen receptor system of gilt uterus. Partial purification of component B labeled with [^{14}C]-iodoacetic acid. *J. Biochem.* **89**, 1829–1837.
5) Fukai, F. and Murayama, A. (1984) Association and dissociation of estrogen receptor with estrogen receptor-binding factors is regulated by Mg^{2+}. *J. Biochem.* **95**, 1227–1230.
6) Fukai, F. and Murayama, A. (1983) Effect of molybdate on the reactivity of basic estrogen receptor molecule with estrogen receptor-binding factors. 56th Congress of the Biochemical Society of Japan, October, Fukuoka (Abstract: *Seikagaku* **55**, No. 8, 840 (1983) (in Japanese)).
7) Hazato, T. and Murayama, A. (1981) Protection of steroid hormone receptors by protease inhibitors. *Biochem. Biophys. Res. Commun.* **98**, 488–493.
8) Jensen, E.V. and DeSombre, E.R. (1973) Estrogen-receptor interaction. *Science* **182**, 126–134.
9) Joab, I., Radanyi, C., Renoir, M., Buchou, T., Catelli, M.-G., Binart, N., Mesler, J., and Baulieu, E.-E. (1984) Common non-hormone binding component in non-transformed chick oviduct receptors of four steroid hormones. *Nature* **308**, 850–853.
10) Martin, R.G. and Ames, B.N. (1961) A method for determining the sedimentation behavior of enzymes: Application of protein mixtures. *J. Biol. Chem.* **236**, 1372–1377.
11) Mueller, G.C., Vonderhaar, B., Kim, U.H., and Mahieu, L.M. (1972) Estrogen action: an inroad to cell biology. *Recent Prog. Horm. Res.* **28**, 1–49.
12) Murayama, A. (1985) Molecular organization of the estrogen receptor system.

In "Molecular Mechanism of Steroid Hormone Action," ed. Moudgil, V.K., pp. 1-30, Walter de Gruyter, Berlin, New York.
13) Murayama, A. and Fukai, F. (1981) Characterization of estrogen receptor, estrogen receptor-binding factors and a protease in the nuclei of gilt uterus. *J. Biochem.* **90**, 823–832.
14) Murayama, A. and Fukai, F. (1982) Inhibition of the translocation of estrogen receptor from the cytoplasm into the nucleus by estrogen receptor-binding factors. *J. Biochem.* **92**, 2039–2042.
15) Murayama, A. and Fukai, F. (1983) Studies on the molecular mechanism of the translocation of estrogen receptor from the cytoplasm into the nucleus. *J. Biochem.* **94**, 511–519.
16) Murayama, A. and Fukai, F. (1983) The strong hydrophobic domain of the activated estrogen receptor molecule. *FEBS Lett.* **158**, 255–258.
17) Murayama, A. and Fukai, F. (1984) Regulation of the reactivity of estrogen receptor by phosphate. 57th Congress of the Biochemical Society of Japan, October, Tokyo (Abstract: *Seikagaku* **56**, No. 8, 838 (1984) (in Japanese)).
18) Murayama, A., Fukai, F., Hazato, T., and Yamamoto, T. (1980) Estrogen receptor of cow uterus. I. Characterization of native and proteolyzed "4S" estrogen receptors. *J. Biochem.* **88**, 955–961.
19) Murayama, A., Fukai, F., Hazato, T., and Yamamoto, T. (1980) Estrogen receptor of cow uterus. II. Characterization of a cytoplasmic factor which binds with native "4S" estrogen receptor to give "8S" estrogen receptor. *J. Biochem.* **88**, 963–968.
20) Murayama, A., Fukai, F., and Murachi, T. (1984) Action of calpain on the basic estrogen receptor molecule of porcine uterus. *J. Biochem.* **95**, 1697–1704.
21) Murayama, A., Fukai, F., and Yamamoto, T. (1980) Estrogen receptor of cow uterus. III. Molecular constitution of estrogen receptor-binding factors. *J. Biochem.* **88**, 969–976.
22) Murayama, A., Fukai, F., and Yamamoto, T. (1980) Progesterone receptor system of hen oviduct. *J. Biochem.* **88**, 1305–1315.
23) Murayama, A., Fukai, F., and Yamamoto, T. (1980) Disorganization and *in vitro* assembly of the constituents of the cytoplasmic estrogen receptor system of cow uterus. *J. Biochem.* **88**, 1457–1466.
24) Schrader, W.T., Kuhn, R.W., and O'Malley, B.W. (1977) Progesterone-binding components of chick oviduct. Receptor B subunit protein purified to apparent homogeneity from laying hen oviducts. *J. Biol. Chem.* **252**, 299–307.
25) Sica, V., Nola, E., Puca, G.A., and Bresciani, F. (1976) Estrogen binding proteins of calf uterus. Inhibition of aggregation and dissociation of receptor by chemical perturbation with NaSCN. *Biochemistry* **15**, 1915–1923.
26) Siegel, L.M. and Monty, K.J. (1966) Determination of molecular weights and

frictional ratios of proteins in impure states by use of gel filtration and density gradient centrifugation. Application to crude sulfite and hydroxylamine reductases. *Biochim. Biophys. Acta* **112**, 346–362.

27) Stancel, G.M., Leung, K.M.T., and Gorski, J. (1973) Estrogen receptors in the rat uterus. Relationship between cytoplasmic and nuclear forms of the estrogen binding protein. *Biochemistry* **12**, 2137–2141.

28) Suda, H., Aoyagi, T., Hamada, M., Takeuchi, T., and Umezawa, H. (1972) Antipain, a new protease inhibitor from *Actinomycetes*. *J. Antibiot.* **25**, 263–266.

29) Vedeckis, V.W. (1983) Subunit dissociation as a possible mechanism of glucocorticoid receptor activation. *Biochemistry* **22**, 1983–1989.

30) Tata, J.R. (1974) Isolation of nuclei from liver and other tissues. In "Methods in Enzymology," Vol. 31, eds. Fleisher, F. and Packer, L., pp. 253–262, Academic Press, New York.

8

Glucocorticoid Hormone: Mechanisms of Action, ed. Sakamoto, Y. and Isohashi, F., pp. 175–193, Japan Sci. Soc. Press, Tokyo/Springer-Verlag, Berlin (1986)

STIMULATORY AND INHIBITORY PATHWAYS FOR STEROID RECEPTOR TO REGULATE GENE EXPRESSION IN HORMONE-RESPONSIVE CELLS

BUNZO SATO,[*1] YASUKO NISHIZAWA,[*1]
YOSHIAKI MAEDA,[*1] KEIZO NOMA,[*1]
SUSUMU KISHIMOTO,[*1] and KEISHI MATSUMOTO[*2]

The Third Department of Internal Medicine[*1] *and The Second Department of Pathology,*[*2] *Osaka University Hospital, Osaka 553, Japan*

The action mechanism of steroid hormones has been extensively studied since the "receptor" was found in the target cells (*16*). After penetrating the cells, probably by passive diffusion (*30*), the hormonal steroid is associated with a special cytosolic protein, the receptor. The complex formation with the corresponding ligand makes the receptor more nucleophilic, resulting in translocation of the complex into nuclei (*15*). These steroid-receptor complexes, in turn, can act on a specific genome as a direct gene stimulator *via* a mechanism which has not yet been identified (*31*). Although this intracellular pathway by which the steroid receptor regulates gene expression has been characterized in more detail than those of other gene regulators including peptide hormones, thyroxine, and certain growth factors, our current knowledge on the steroid hormone action mechanism is far from complete. For instance, the native form of the cytosolic receptor (even its molecular weight) remains controversial. Upon binding to the hormonal steroid, conformational changes of the receptor molecules are induced. These can be shown by enhanced stability of the receptor functions (*43*). The molecular nature of this steroid binding-induced stabilization is also undetermined. The most critical step for conversion of a receptor into the biologically active form seems to be "receptor activation" which is defined as the acquisition of nuclear binding activity. Despite extensive studies by many groups including our laboratory (*3*, *10*, *41*, *43*, *48*), the molecular mechanism of this receptor activation process has not yet been fully elucidated. The possibility has been proposed, but not proved, that the nuclear steroid receptor complexes stimulate gene expression in a fashion analogous to gene regulators in prokaryocytes which can bind to a specific DNA sequence in the promoter region of genes under their control, thereby changing the rate of transcription (*9*, *17*, *32*). Finally, the mechanism, which turns off the transcription of genes once stimulated by the receptor-steroid complexes, is entirely unknown. In view of these unresolved questions on the steroid hormone action mechanism, re-evaluation of the many proposed theories in this interesting field of molecular biology would be valuable.

Estrogen receptor (ER) systems have been the most extensively analyzed of the steroid receptor systems since estrogen is known to affect important cellular functions in target cells, such as proliferation, differentiation, and transformation. In addition, the ER molecule has been observed to be relatively stable in terms of steroid binding ability, and this stability has allowed its detailed characterization. The conformational change of ER in response to activation procedures was identified first (*15*). Although it is still questionable whether or not the conversion of ER from 4 S to 5 S state, as determined by the high salt sucrose density gradient, is obligatory for estrogen action (*15*, *34*), this observation clearly demonstrated that simple procedures such as heating result in a marked conformational change of the ER molecule with simultaneous alteration in its function. In an attempt to identify the specific binding site(s) on chromatin for ER, intranuclear events in the pathway to affect transcriptional activity were investigated, and a report on the processing of estrogen-ER complexes, which is marked by the loss of nuclear binding without the reappearance of ER in the cytosol, was interesting (*11*). Correlations between certain estrogen responses and processing suggest that this may be an important step in estrogen action.

Based on these observations, many hypotheses have been proposed in connection with the relationship between the molecular status of ER and estrogen action. Recent studies on kinetic alteration in ligand-ER interaction elicited by *in vitro* manipulations raised the possibility that the conversion of

ER from low-affinity to high-affinity state is a prerequisite for ER stimulation of gene expression (*33*, *48*). A careful survey of these reported results, however, reveals that they are based on experiments involving a relatively short-term exposure of non-estrogenized cells to estrogen. Prolonged treatment with estrogen might be expected to alter the cellular sensitivity to this substance by a yet unidentified mechanism. Extensive stimulation of target cells with peptide hormones is known to cause a decrease in the number of binding sites (so-called down-regulation), resulting in the resistance of cells to additional hormonal stimuli (*4*). A clarification of this mechanism in steroid hormone action, if any, would be important in relation to protection of the target cell from excessive estrogenic stimulation. This could also provide information on the molecular status of biologically active or inactive ER. In this regard, triethylphenylene compounds, so-called antiestrogens, have been introduced as a useful tool in the field of steroid receptor research. Many reported results obtained using these interesting compounds have hinted at an explanation of their antiestrogenic properties. However, their action mechanism remains to be proved (*46*).

This unresolved status of the steroid hormone action mechanism makes the examination of ER characteristics under various hormonal conditions advisable. Our approach to understanding the molecular mechanism of these actions has been to monitor the alterations in estrogen sensitivity of target cells in response to various hormonal conditions and to establish a model system by which to analyze how hormonal signals can be turned off. We therefore summarize our recent results in hopes of obtaining some clue to clarify the intracellular regulatory mechanism(s) for ER which controls the cellular functions.

I. QUANTITATIVE AND QUALITATIVE ANALYSES OF ER UNDER VARIOUS HORMONAL CONDITIONS

A special mechanism has been thought to protect the target cell against overstimulation with hormones. An event similar to down-regulation in the peptide hormone systems was recently reported in the glucocorticoid system (*6*). The continuous exposure of the target cell to excessive glucocorticoid resulted in a loss of cellular glucocorticoid binding sites; however, this observation and its biological significance require further confirmation. Accordingly, much more uncertainty surrounds this protective mechanism in the steroid hormone

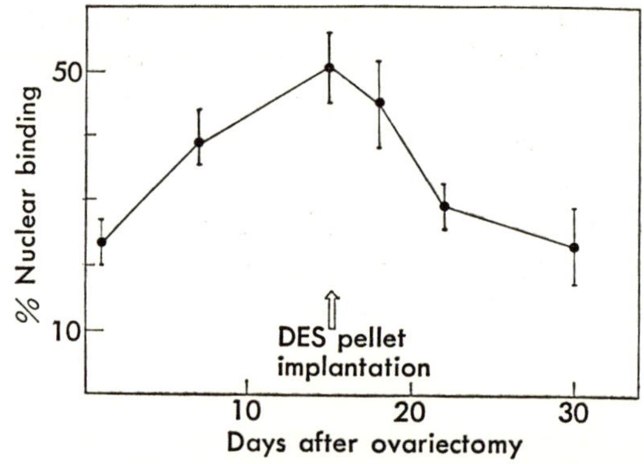

Fig. 1. Cell-free activation of uterine ER in non-estrogenized and estrogenized rats. Adult rats were ovariectomized on day 0 and sacrificed at the indicated period to obtain the uterine cytosol in 4 volumes of 0.01 M tris, 2 mM mercapto-ethanol, 1.5 mM EDTA, and pH 7.4. After incubation at 0°C for 2 hr with 4 nM [^{3}H]-E_2 $\pm$0.4 μM unlabeled E_2, these complexes were heated at 25°C for 30 min, followed by incubation with isolated nuclei to measure their nuclear binding activity (*43*). To estrogenize the rats, the DES pellet was implanted on day 15. The uterine cytosol was obtained from these rats and treated with dextran-coated charcoal, and the treated cytosol was then subjected to nuclear binding assay. The data (mean$\pm$S.E.) are expressed as % nuclear binding (*43*).

system in comparison with that in the peptide hormone system. Since ER functions are known to be regulated by many environmental factors (*42*), we first compared the interaction of ER with the cytoplasmic modulator between estrogen-stimulated and -unstimulated target cells (*27*). During studies on the cell-free ER activation process, estrogenization of rats resulted in a diminished nuclear binding ability of [^{3}H]-E_2 ER complexes in response to heat activation (Fig. 1). The rapid loss of the nuclear binding ability of once-activated ER was also induced by prolonged exposure to relatively high (20–30°C) temperatures of the uterine cytosol from adult rats, but not from non-estrogenized immature rats. These findings can be explained by the presence of a macromolecular modulator in the adult rat uteri which recognizes the activated ER but not the glucocorticoid receptor (GR). The interaction of this modulator with activated ER resulted in a rapid shift of the ER molecule into the large aggregated form with simultaneous loss of its nuclear binding ability. Interestingly, this reaction was accelerated by the introduction of KCl (0.15–0.4 M) into the buffer system. Additionally, this unique modulator,

TABLE I
Comparison of ER Aggregation Factor with So-called Macromolecular Inhibitor

	ER aggregation factor	Macromolecular inhibitor
Molecular size	Macromolecular	Macromolecular
Function	Decreased nuclear binding	Decreased nuclear binding
Receptor specificity	ER only	All steroid receptors
Localization	Rat uteri	Many steroid target cells
Reversibility	Irreversible	Reversible
KCl effect	Enhanced	
RNase treatment	Resistant	Sensitive (?)
Temperature (25°C)	Enhanced	

The above characteristics were determined in experiments described in ref. *27*.

which we call the ER aggregation factor, was even absent in the liver from estrogenized female adult rats. The macromolecular inhibitor is known to be involved in the process of translocation of activated steroid receptor into nuclei (*5*). The properties of these modulators, which are summarized in Table I, would indicate that the ER aggregation factor newly identified is entirely different from the so-called macromolecular inhibitor.

The presence of the modulator, whose synthesis may be under estrogen control and whose role may be related to the reduced translocation of activated ER, prompted us to study ER dynamics in non-estrogenized and pre-estrogenized rat uteri. Difficulty was anticipated in using rat uterine cytosol in the cell-free system because of heterogeneity of uterine cell populations and the contamination of extracellular ground substances. We therefore employed the *in vivo* estradiol-17β (E_2) injection method to monitor the receptor dynamics which could be determined by exchange assays (*1*, *5*). To examine this point, rats were pre-estrogenized with a one-week subcutaneous implantation of one diethylstilbestrol (DES) pellet (20 mg-pellet composed of cholesterol and DES (9:1, W/W)) (*34*). This pre-estrogenization procedure resulted in no significant loss of the cytoplasmic or cellular ER content, indicating that the down-regulation mechanism does not work in the ER system (*2*). When a pharmacological dose (0.1 μg/g body weight) of E_2 was subcutaneously injected to these rats, a significant increase in nuclear ER content was observed 1 hr after injection. However, the maximum level of nuclear ER in pre-estrogenized rat uteri was approximately half of that obtained by E_2 injection into estrogen-deprived rats (Fig. 2). The possibility should be considered that this

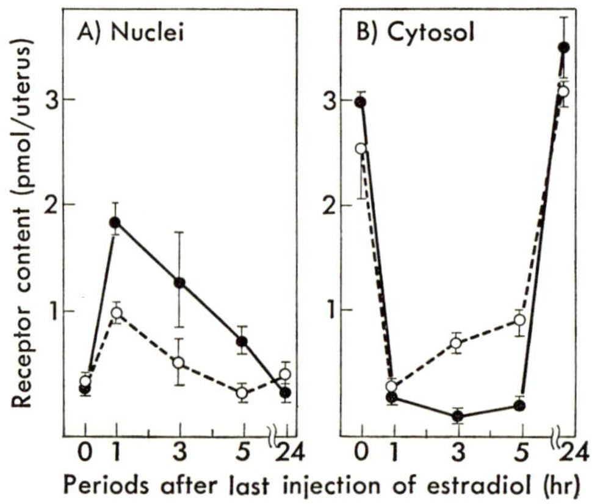

Fig. 2. ER dynamics in pre-estrogenized or non-estrogenized rat uteri after a single injection of E_2. Rats received an ovariectomy alone (●) or an ovariectomy plus a DES pellet implantation (○) 12 days before the experiments. E_2 was injected and the nuclear (A) as well as cytosol (B) receptor contents were monitored at the periods indicated after injection (mean±S.E.).

diminished nuclear accumulation of ER in response to E_2 injection is due to experimental artifacts during *in vitro* exchange assays. However, protease-, nuclease- or phosphatase-mediated degradation of nuclear ER in pre-estrogenized rat uteri during assays seems unlikely, as shown by the use of various kinds of inhibitor (*40*). Moreover, it has been proved that this is not due to the progesterone-induced alteration in ER dynamics reported by Evans and Leavitt (*8*), since adrenalectomy and ovariectomy did not affect estrogenization-dependent diminution of nuclear ER accumulation. Therefore, this phenomenon reflects naturally occurring intracellular events. The current theory on steroid hormone action mechanism requires consideration of two critical steps for nonactivated ER to reach acceptor sites on the chromatin, namely, receptor activation and translocation of activated ER into nuclei. We previously reported that the receptor activation step is principally regulated by the so-called low molecular weight inhibitor (*41*, *43*). To date, there is no evidence of a difference in the content of the low molecular weight inhibitor between estrogenized and non-estrogenized target tissues. The partially purified ER obtained by DEAE-cellulose chromatography, hydroxylapatite chromatography, and $(NH_4)_2SO_4$ fractionation (*38*) showed a similar degree of nuclear binding ability in both estrogenized and non-estrogenized uteri, indicating that estrogenization-induced modification of the ER molecule with a resultant loss of its nuclear binding ability does not occur (unpublished observation). Therefore, this diminished nuclear accumulation can be explained as an event beyond the activation step.

Another interesting alteration in ER dynamics emerging from this experiment is the short-lived accumulation of ER in the nuclei. After the temporal increase in nuclear ER levels described above, the nuclear ER content in DES-prestimulated rats returned to the preinjection value 3 hr after E_2 administration. Non-estrogenized rats, however, showed a relatively prolonged retention of ER in the nuclei and the value 5 hr after injection remained significantly higher than the preinjection value. The cytosol receptor contents in both groups of animals reached a nadir 1 hr after injection and then gradually increased, although the replenishment of the cytosol ER was observed to occur more rapidly in estrogenized rats than in non-estrogenized rats. The biological significance of a short-lived nuclear accumulation of steroid receptor is still a matter of conjecture. Multiple injections of progesterone to nonovariectomized rabbits have been reported to render attenuated processing of receptors (*14*). These results are interpreted as an indication that accelerated processing is related to enhanced biological responses in terms of uteroglobin secretion. However, this conclusion cannot be drawn if a marker with a very short biological half life is not used to monitor the sensitivity of the target cells. In other words, it is highly possible that an apparent good response to hormonal stimuli after multiple injections is simply due to an accumulated effect rather than to attenuated receptor processing.

The biological role of receptor processing was also deduced from experiments involving the use of so-called antiestrogens (*7*, *12*). The lack of processing of ER associated with antiestrogen was reported. Although this event was said to be related to their antiestrogenic properties, a single injection of antiestrogen is known to elicit an estrogenic response in many target cells (see below). In addition, this apparent lack of processing of ER-antiestrogen complexes has been theorized to emerge simply from the prolonged retention of triethylphenylene compounds in animals (*18*). In connection with this altered receptor dynamics, we previously reported that E_2 injection into C_3H_{Bi} mice resulted in a shorter-lived and smaller amount of ER accumulation in the nuclei of Leydig cell in comparison with that in BALB/c mice (*44*). Interestingly, Leydig cell in C_3H_{Bi} mice does not respond to estrogen administration while that in BALB/c mice responds well leading to ultimate malignant transformation (*45*). In view of these observations, we feel tempted to speculate that this altered ER dynamics elicited by prolonged treatment with a large dose of estrogen might be related to the acquisition of refractoriness to additional estrogenic stimuli in pre-estrogenized target cells. The molecular mechanism of alteration in ER dynamics remains unknown. The ER aggregation factor, described in the previous section, might be related to the diminished nuclear accumulation of ER. Estrogen treatment might induce chromatin-associated phosphatase which has been proposed to have an important role in ER processing (*2*). Further studies are required to examine these possibilities.

On the biological role of nuclear ER, accumulated evidence shows that the presence of the steroid receptor in the nuclei is not enough to trigger the hormonal action. We recently observed that ER in one of the mouse Leydig cell tumor lines was exclusively localized in the nuclear fraction even in the absence of estrogenic stimuli (*23*, *35*). The biological significance of nuclear localization of ER was therefore addressed, and the presence of unoccupied nuclear ER was shown to be insufficient to trigger estrogen actions; the complex formation of unoccupied nuclear ER with estrogen is a prerequisite for full estrogen action in tumor growth. Determination of the presence of nuclear ER in a cell which is not exposed to estrogen has been reported using different systems (*22*, *49*). These observations suggest that ER is associated with the nuclear component(s) which is the site for nonstimulatory or even inhibitory effects in estrogen action. To uncover the biological significance of nuclear accumulation and dynamics of ER, therefore, a marker is required which faithfully and rapidly reflects estrogenic stimuli.

II. ESTROGEN-INDUCED BIOLOGICAL EVENTS AND THEIR APPLICATION TO ANALYSES OF FEEDBACK MECHANISM AT THE TARGET CELL LEVEL

The effects of steroid hormones on the target cells have been monitored by many markers. Since rat uteri have played a key role in elucidating the receptor mechanism, markers to measure the biological activity of ER are needed. Some functional proteins such as progesterone receptor (*26*) and ornithine decarboxylase (*20*) have been used as markers of estrogen action, however, these do not seem satisfactory to discuss the transcriptional regulation by estrogens since the possibility of a post-transcriptional regulation mechanism cannot be completely eliminated. The so-called induced protein (IP), first described by Notides and Gorski (*29*), is also inappropriate as a marker to examine ER activity because of its minor increase (100–400%) with estrogen administration and its lack of target specificity (*13*). Other gene products are needed to monitor estrogenic stimuli under various hormonal conditions.

Our approach to the determination of new products was to isolate mRNAs from the rat uteri and to analyze their cell-free translation products using rabbit reticulocyte lysate (*28*). Poly (A)-rich RNAs were extracted from the uteri of rats which had been estrogenized by three injections of a pharmacological dose (0.1 μg/g body weight) of E_2 at 24 hr intervals. These were

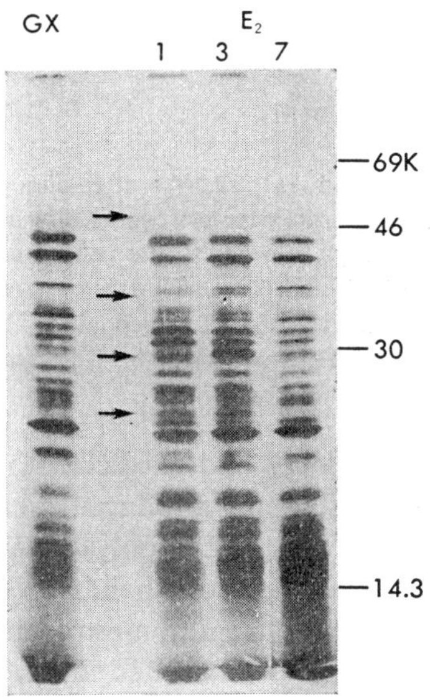

Fig. 3. Appearance and disappearance of estrogen-induced mRNAs in rat uteri by multiple E_2 injections. Rats were ovariectomized and used for experiments no earlier than 4 weeks after surgery. E_2 was injected subcutaneously at a dose of 0.1 μg/g body weight at 24 hr intervals. The animals were sacrificed 24 hr after the last injection and subjected to extraction and purification to uterine mRNAs (*28*). The cell-free translation products directed by these mRNAs were analyzed using SDS polyacrylamide gel electrophoresis and fluorography. GX: ovariectomized alone; E_2 1, 3, 7: E_2 injected for 1, 3, and 7 days. Arrows indicate the products translated by estrogen-induced mRNAs.

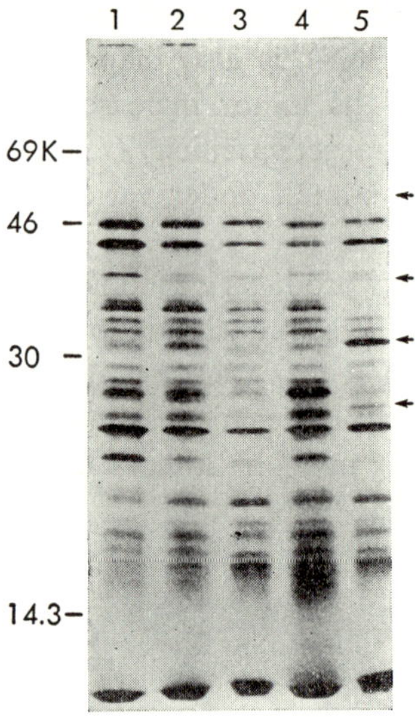

Fig. 4. Rapid disappearance of estrogen-induced mRNAs after discontinuation of estrogen administration. Ovariectomized rats were estrogenized by subcutaneous implantation of a DES pellet for 7 days. On day 8, the pellet was removed and the animals were killed on the day indicated. Uterine mRNAs were analyzed as described in the legend of Fig. 3. Lane 1: ovariectomized alone; lanes 2, 3, and 4: 1, 3, and 7 days after removal of DES pellet; lane 5: estrogenized (no removal).

purified using oligo-dT-cellulose chromatography and subjected to *in vitro* cell-free translation assay. When these translational products, which were labeled with [^{35}S]-methionine, were analyzed with SDS polyacrylamide gel electrophoresis, estrogenization appeared as at least four bands on the fluorography which could not be found in the sample from non-estrogenized rats (Fig. 3). Judged by their mobility, the proteins translated by E_2-induced mRNAs had molecular weight of 26,500, 32,000, 38,000, and 53,000, respectively. The most prominent of these bands was a protein with the molecular weight of 32,000. These bands can be induced by *in vivo* DES administration to rats, suggesting that mRNA biosynthesis is under the stringent control of estrogen. Discontinuation of the estrogenization caused the rapid disappearance of these estrogen-induced mRNAs, resulting in the almost complete vanishing of estrogen-induced bands by 3 days after estrogen withdrawal (Fig. 4), and indicating the relatively short biological half life of these mRNAs.

Accordingly, we considered these mRNAs to be markers suitable for elucidation of the "on and off" mechanism of estrogen action.

Using these markers, the effect of daily injection of a pharmacological dose (0.1 μg/g body weight) of E_2 on the rat uteri was studied. As shown in Fig. 3, these estrogen-induced mRNAs were clearly identified 24 hr after the initial injection. Three injections were found to markedly enhance the intensity of these bands. Surprisingly, repeated injections for 7 days resulted in the disappearance of these estrogen-induced bands without detectable alteration in the other fluorographic bands. There are several possible reasons for this: i) that the accelerated E_2 metabolism induced by repeated E_2 injection causes the conversion of E_2 into the inactive metabolites before its binding to uterine ER; ii) that E_2 treatment induces RNase, thereby destroying estrogen-induced mRNAs; iii) that prolonged E_2 treatment induces cell differentiation from estrogen-sensitive to -resistant cells; and iv) that uterine cells contain both a stimulatory and an inhibitory pathway for estrogen actions, and that overstimulation with estrogen antagonizes the actions *via* the latter pathway. When radioactive E_2 was injected to fully estrogenized rats, the sufficient amount of E_2 was found to be associated with ER (see below). Further, the finding that multiple E_2 injection limits the disappearance of estrogen-induced bands with no detectable alteration in other bands suggests that RNase problems are not the cause. Although no direct evidence is yet available, irreversible differentiation into estrogen-resistant cells does not seem to be induced by multiple E_2 injections, since no difference in mRNA-directed translation products was detected between non-estrogenized and estrogenized uteri except those from estrogen-induced mRNAs. Collectively, we favor the last possibility, that both stimulatory and inhibitory pathways exist in many estrogen target cells.

We next asked whether or not estrogen-induced altered ER dynamics are related to the antagonism of mRNA biosynthesis with repeated injections of a pharmacological dose of E_2. E_2 was injected daily into ovariectomized rats for 7 days and the nuclear ER accumulation elicited by additional E_2 injection was assayed on day 8 (Fig. 5). Pre-estrogenization by repeated E_2 injection markedly lowered the maximum level of nuclear ER and shortened its nuclear retention time when compared to ER dynamics in non-estrogenized rat uteri. These results combined with those shown in Figs. 2 and 4 suggest that prolonged nuclear retention of ER is important in triggering estrogen action, and that estrogenization-induced alteration in its nuclear accumula-

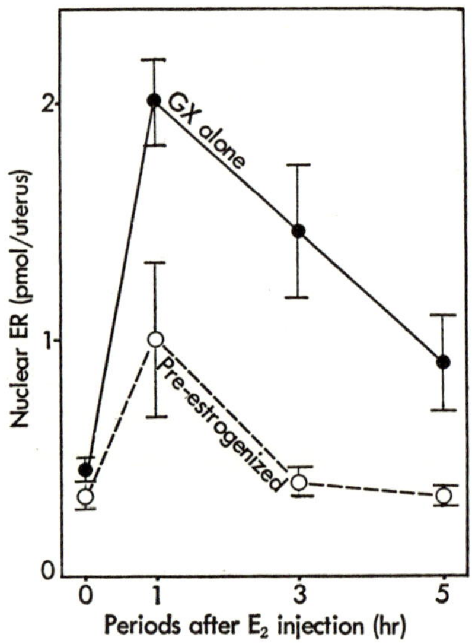

Fig. 5. Nuclear accumulation pattern of ER in pre-estrogenized and non-estrogenized rat uteri by E_2 injection. Daily injections of E_2 (0.1 μg/g body weight) into ovariectomized rats were given for 7 days (○). As a control, ovariectomized rats without E_2 injection were employed (●). On day 8, both groups of animals were injected with E_2 (0.1 μg/g body weight) and sacrificed at different periods of time thereafter. The nuclear receptor content was measured by exchange assay.

tion pattern is related to the mechanism of estrogenization-induced desensitization to estrogen in the target cells.

Based upon these observations, we would be tempted to speculate that there are two different pathways in estrogen target cells, one for stimulatory and the other for inhibitory action. This idea is supported by the interesting observation that prolonged treatment with a large dose of estrogen inhibits the growth of estrogen-responsive tumors (*21*). Stimulation with a sustained high level of estrogen was also reported to result in diminished induction of progesterone receptor in rat uteri (*24*). To obtain conclusive evidence for the two-different pathway concept, however, a more sophisticated system is needed, and we tried to identify the target cell which contains the single pathway.

III. IDENTIFICATION OF MOUSE LEYDIG CELL TUMOR CONTAINING SINGLE PATHWAY FOR ER

Our recent observations revealed that mouse Leydig cell tumor lines are effective in elucidating the estrogen action mechanism. Chronic estrogenization

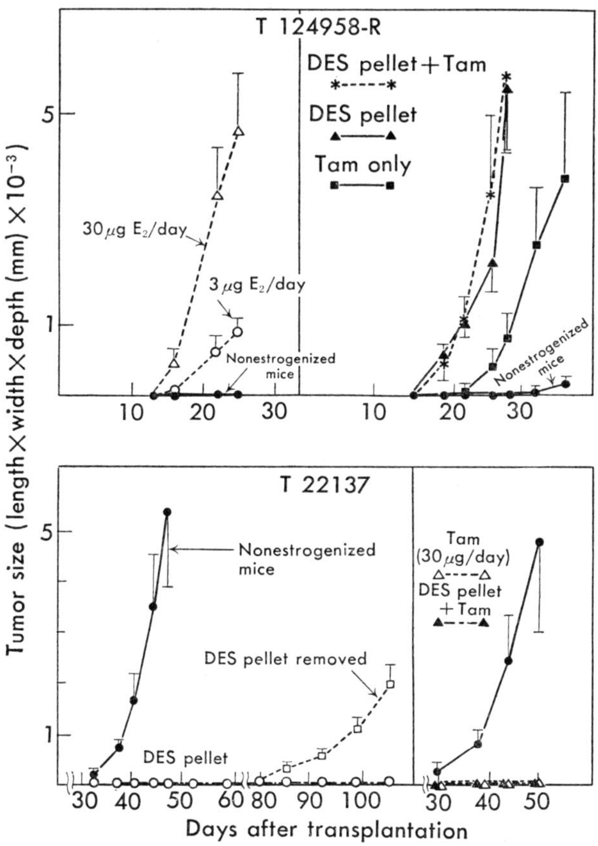

Fig. 6. Estrogen- and/or antiestrogen-dependent alteration in the growth of mouse Leydig cell tumor lines. T 124958-R (upper panel) or T 22137 (lower panel) was transplanted into BALB/c mice. These mice were treated with E_2 (3 μg/mouse or 30 μg/mouse), E_2 pellet, DES pellet, or tamoxifen (30 μg/mouse) $\pm$ DES pellet, and the tumor size was measured.

develops the transplantable tumor in a strain-specific manner. Some tumors have been found to contain unique estrogen binding components in the cytoplasm (*37*) and in the nuclei (*23*, *35*); a couple of tumor lines have also been shown to be markedly influenced by estrogenization of host mice in terms of their growth (*23*). We attempted to establish a tumor line with a growth either stimulated or inhibited by estrogen administration (*36*). First, we selected a new tumor line whose growth was markedly accelerated without any negative effect by estrogenization of host mice. After repeated selection by passing from 8 to 10 generations, a new line (T 124958-R) showing marked enhancement of tumor growth in response to estrogenization was established. The other line used in this experiment was T 22137, which was known to be inhibitorily influenced by estrogenic stimuli with respect to its growth. If these tumor lines contain only a single pathway for ER, it is highly possible that triethylphenylene compounds (so-called antiestrogen) act in these cells in a fashion analogous to estrogen. Before further discussion, a brief comment on the effect of tamoxifen (the most widely used antiestrogen) on mice is appropriate. Tamoxifen has been recognized as a good reagent to study the "on and off" mechanism of estrogen action, and has a species-dependent difference in action. For instance, its administration to mice has been reported to show estrogenic, but not antiestrogenic, influence to the uteri (*47*). However, estrogen-induced enhancement of the growth of a mouse tumor has been antagonized by tamoxifen, indicating that it can act as an antiestrogen even in the mouse system (*25*). We examined tamoxifen and/or estrogen effect on the growth of mouse Leydig cell tumor lines. The daily injection of a relatively low dose (3 μg/mouse) of E_2 into T 124958-R bearing mice enhanced tumor growth (Fig. 6). Administration of a large E_2 dose resulted in even further tumor growth enhancement, suggesting that a pharmacological dose of estrogen cannot antagonize the estrogen-induced tumor growth, a pattern different from that in many estrogen-responsive tumors (*21*).

Interesting results have been obtained in experiments involving tamoxifen: its 30 μg/mouse/day injection into host mice resulted in an acceleration of the growth of T 124958-R; estrogen-induced enhancement of the tumor growth was not antagonized by simultaneous tamoxifen administration; this tumor cell therefore seems to contain only the stimulatory pathway for both estrogen and antiestrogen. On the other hand, estrogenization of mice with a DES or E_2 pellet resulted in inhibition of the growth of T 22137. Removal of the DES pellet after two-months of estrogenization allowed the tumor again to grow, suggesting that estrogen did not kill the tumor cell but locked it into a nonproliferative state. We again addressed the possibility that estrogen-induced inhibition of tumor growth can be antagonized by tamoxifen administration; however, the results clearly showed that tamoxifen inhibits the growth of T 22137 irrespective of the estrogen status of host mice. Thus, it is reasonable to conclude that T 22137 contains only the inhibitory pathway for estrogen or tamoxifen. To gain insight into the molecular basis of the difference in estrogen action between T 124958-R and T 22137, the level of both cytoplasmic ER and antiestrogen binding sites was measured; results indicated no clear difference between the two tumor lines (*36*). Measurement of the dissociation rate constant of [^{3}H]-E_2-ER complexes, which has been reported to be able to differentiate the stimulatory and inhibitory forms of ER (*33*), failed to demonstrate a clear difference between the two tumor lines. At this preliminary stage, therefore, our concept is that the difference in estrogen action must be at the post-receptor level, although there is still a possibility of a subtle difference in ER molecule between the two tumor lines which causes the opposite estrogen action. Detailed analysis of ER molecules and their interaction with chromatin are a future project to be clarified. Estrogen-induced or -inhibited mRNAs as well as proteins must be identified in these tumor lines to understand the dual effects of estrogenic compounds. We recently successfully demonstrated that T 124958-R secretes glycosylated protein with a molecular weight of 34,000 in response to the estrogenization of host mice (*39*). The effect of tamoxifen on this estrogen-induced secretory protein is now under study in our laboratory.

SUMMARY

In this review article, we have documented that estrogen actions are sophisticatedly regulated at the target cell level, especially in terms of their inhibitory effects on the expression of estrogen-responsive genes. This event, which we would like to call the "intracellular feedback mechanism of estrogen action," has a biological role, the molecular basis of which remains obscure. The possibility that glucocorticoid as well as androgen action is also regulated by a mechanism similar to that in estrogen action should be examined. In this regard, it is well to remember that an exposure to glucocorticoid causes cell death in some target cells and enhanced cell proliferation in other cells. Elucidation of the molecular mechanism of this dual effect of glucocorticoid would also be important in the area of clinical medicine, since the beneficial effects of administration of a large amount of glucocorticoid to some patients, so-called pulse-therapy, are well recognized. These interesting projects await further experimental analysis.

Acknowledgments
This work was supported in part by grants-in-aid from the Ministry of Education, Science and Culture of Japan, the Japanese Cancer Research Foundation, Naito for 1979, the Hisamitsu Cancer Research Fund and the Hirai Cancer Research Fund. We are also grateful to ICI Pharma (Osaka, Japan) for providing us with tamoxifen.

REFERENCES

1) Anderson, J., Clark, J.H., and Peck, E.J., Jr. (1972) Oestrogen and nuclear binding sites. *Biochem. J.* **126**, 561–567.
2) Auricchio, F., Migliaccio, A., and Rotondi, A. (1981) Inactivation of oestrogen receptor *in vitro* by nuclear dephosphorylation. *Biochem. J.* **194**, 569–574.
3) Cake, M.H., Goidl, J.A., Pearchman, L.G., and Litwack, G. (1976) Involvement of a low molecular weight component(s) in the mechanism of action of the glucocorticoid receptor. *Biochem. Biophys. Res. Commun.* **71**, 45–52.
4) Catt, K.J. and Dufaw, M.L. (1977) Peptide hormone receptors. *Annu. Rev. Physiol.* **39**, 529–557.
5) Chamness, G.C., Jennings, A.W., and McGuire, W.L. (1974) Estrogen receptor binding to isolated nuclei. A nonsaturable process. *Biochemistry* **13**, 327–331.
6) Cidlowski, J.A. and Cidlowski, N.B. (1981) Regulation of glucocorticoid receptors by glucocorticoids in cultured Hela S_3 cells. *Endocrinology* **109**, 1975–1982.
7) Clark, J.H., Peck, E.J., Jr., and Anderson, J.N. (1974) Oestrogen receptors and antagonism of steroid hormone action. *Nature* **251**, 446–448.
8) Evans, R.W. and Leavitt, W.W. (1980) Progesterone action in hamster uterus: Rapid inhibition of ^{3}H-estradiol retention by the nuclear fraction. *Endocrinology* **107**, 1261–1263.
9) Govindan, M.V., Spiess, E., and Majors, J. (1982) Purified glucocorticoid receptor-hormone complex from rat liver cytosol binds specifically to cloned mouse mammary tumor virus long terminal repeats *in vitro*. *Proc. Natl. Acad. Sci. U.S.A.* **79**, 5157–5161.
10) Gradics, P., Miller, A., Schmidt, T.J., Mittman, D., and Litwack, G. (1984)

Purification of the unactivated glucocorticoid receptor and its subsequent *in vitro* activation. *J. Biol. Chem.* **259**, 3173–3180.

11) Horwitz, K.B. and McGuire, W.L. (1978) Estrogen control of progesterone receptor in human breast cancer: Correlaton with nuclear processing of estrogen receptor. *J. Biol. Chem.* **253**, 6319–6322.

12) Horwitz, K.B. and McGuire, W.L. (1978) Nuclear mechanism of estrogen action. Effects of estradiol and anti-estrogens on estrogen receptors and nuclear receptor processing. *J. Biol. Chem.* **253**, 8185–8191.

13) Iacobelli, S., Kind, R.J.B., and Vokaer, G. (1977) Antibody to estrogen-induced protein (IP) and quantitation of the protein in rat uteri by a radioimmunoassay. *Biochem. Biophys. Res. Commun.* **76**, 1230–1237.

14) Isotalo, H., Isomaa, V., and Jänn, O. (1981) Replenishment and properties of cytosol progesterone receptor in rabbit uterus after multiple progesterone administrations. *Endocrinology* **108**, 868–873.

15) Jensen, E.V. and DeSombre, E.R. (1971) Mechanism of action of the female sex hormones. *Annu. Rev. Biochem.* **41**, 203–230.

16) Jensen, E.V. and Jacobson, E.P. (1962) Basic guides to the mechanism of estrogen action. *Recent Prog. Horm. Res.* **18**, 387–417.

17) Jost, J.-P., Seldran, M., and Geiser, M. (1984) Preferential binding of estrogen-receptor complex to a region containing the estrogen-dependent hypomethylation site preceding the chicken vitellogenin II gene. *Proc. Natl. Acad. Sci. U.S.A.* **81**, 429–433.

18) Katzenellenbogen, B.S., Katzenellenbogen, J.A., Ferguson, E.R., and Krauthammer, N. (1978) Anti-estrogen interaction with uterine estrogen receptors. Studies with a radiolabeled anti-estrogen (C1-628). *J. Biol. Chem.* **253**, 697–707.

19) Katzenellenbogen, J.A., Johnson, H.J., and Carlson, K E. (1973) Studies on the uterine, cytoplasmic estrogen binding protein. Thermal stability and ligand dissociation rate. An assay of empty and filled sites by exchange. *Biochemistry* **12**, 4092–4099.

20) Kaye, A.M., Icekson, I., and Lindner, H.R. (1971) Stimulation by estrogens of ornithine and S-adenosylmethionine decarboxylases in the immature rat uterus. *Biochim. Biophys. Acta* **252**, 150–159.

21) Kennedy, B.J. (1970) Hormone therapy in cancer. *Geriatrics* **25**, 106–112.

22) King, W.J. and Greene, G.L. (1984) Monoclonal antibodies localize oestrogen receptor in the nuclei of target cells. *Nature* **307**, 745–747.

23) Maeda, Y., Sato, B., Noma, K., Kishimoto, S., Koizumi, K., Aono, T., and Matsumoto, K. (1983) Characterization of a unique nuclear estrogen-binding component in an estrogen-responsive mouse Leydig cell tumor. *Cancer Res.* **43**, 4091–4097.

24) Manni, A., Baker, R., Arafah, B.M., and Pearson, O.H. (1981) Uterine oestrogen and progesterone receptors in the ovariectomized rat. *J. Endocrinol.* **91**, 281–287.
25) Matsuzawa, A., Mizuno, Y., and Yamamoto, T. (1981) Antitumor effect of the antiestrogen, tamoxifen, on a pregnancy-dependent mouse mammary tumor (TPDMT-4). *Cancer Res.* **41**, 316–324.
26) Milgrom, E., Thi, L., Atgar, M., and Baulieu, E.-E. (1973) Mechanism regulating the concentration and the conformation of progesterone receptor(s) in the uterus. *J. Biol. Chem.* **248**, 6366–6374.
27) Nishizawa, Y., Maeda, Y., Noma, K., Sato, B., Matsumoto, K., and Yamamura, Y. (1981) A modulator which converts activated estrogen receptor to a biologically inactive aggregated form. *Endocrinology* **109**, 1463–1472.
28) Nishizawa, Y., Sato, B., Noma, K., Kishimoto, S., and Matsumoto, K. Stimulatory and inhibitory effects of estradiol on estrogen-induced messenger RNAs in rat uterus. Submitted.
29) Notides, A. and Gorski, J. (1966) Estrogen-induced synthesis of a specific uterine protein. *Proc. Natl. Acad. Sci. U.S.A.* **56**, 230–235.
30) O'Malley, B.W. (1971) Unified hypothesis for early biochemical sequence of events in steroid hormone action. *Metabolism* **20**, 981–988.
31) O'Malley, B.W. and Means, A.R. (1974) Female steroid hormones and target cell nuclei. *Science* **183**, 610–620.
32) Payvar, F., DeFranco, D., Firestone, G.L., Edgar, B., Wrange, Ö., Okret, S., Gustafsson, J.-Å., and Yamamoto, K.R. (1983) Sequence-specific binding of glucocorticoid receptor to MTV DNA at sites within and upstream of the transcribed region. *Cell* **35**, 381–392.
33) Rochefort, H. and Borgna, J.L. (1981) Differences between oestrogen receptor activation by oestrogen and antioestrogen. *Nature* **292**, 257–259.
34) Sato, B., Huseby, R.A., and Samuels, L.T. (1978) Characterization of estrogen receptors in various mouse Leydig cell tumor lines. *Cancer Res.* **38**, 2842–2847.
35) Sato, B., Maeda, Y., Nakao, M., Noma, K., Kishimoto, S., and Matsumoto, K. (1985) Multiple estrogen binding sites in malignant mouse Leydig cell and their role for cell proliferation. *Eur. J. Cancer Clin. Oncol.* **21**, 199–205.
36) Sato, B., Maeda, Y., Nishizawa, Y., Noma, K., Kishimoto, S., and Matsumoto, K. (1984) Effects of tamoxifen on estrogen-induced enhancement or inhibition of tumor growth in mouse Leydig cell tumor lines. *Cancer Res.* **44**, 4386–4393.
37) Sato, B., Maeda, Y., Noma, K., Matsumoto, K., and Yamamura, Y. (1981) Estrogen-binding component in mouse Leydig cell tumor. *In vitro* conversion from non-receptor to receptor-like component. *Endocrinology* **108**, 612–619.
38) Sato, B., Nishizawa, Y., Maeda, Y., Noma, K., Honma, T., and Matsumoto,

K. (1983) Steroid receptor forms and their interaction with cytoplasmic modulators. *J. Steroid Biochem.* **19**, 315–321.

39) Sato, B., Nishizawa, Y., Nakao, M., Noma, K., Kishimoto, S., and Matsumoto, K. (1984) Estrogen induces secretory proteins in transformed mouse Leydig cell *in vivo* but not *in vitro*. *Biochem. Biophys. Res. Commun.* **124**, 277–282.

40) Sato, B., Nishizawa, Y., Noma, K., Kishimoto, S., and Matsumoto, K. (1983) Chronic estrogen treatment causes an alteration in uterine estrogen receptor dynamics of rats. *Biochim. Biophys. Acta* **755**, 412–419.

41) Sato, B., Nishizawa, Y., Noma, K., Matsumoto, K., and Yamamura, Y. (1979) Estrogen-independent nuclear binding of receptor protein of rat uterine cytosol by removal of a low molecular weight inhibitor. *Endocrinology* **104**, 1474–1479.

42) Sato, B., Nishizawa, Y., Noma, K., Nakao, M., Kishimoto, S., and Matsumoto, K. (1985) The important role of cytoplasmic modulators in the pathway for steroid receptor to be converted to the biologically active form. In "Molecular Mechanism of Steroid Hormone Action," ed. Moudgil, G.K., pp. 251-278, Walter de Gruyter, Berlin, New York.

43) Sato, B., Noma, K., Nishizawa, Y., Nakao, K., Matsumoto, K., and Yamamura, Y. (1980) Mechanism of activation of steroid receptors: Involvement of low molecular weight inhibitor in activation of androgen, glucocorticoid, and estrogen receptor systems. *Endocrinology* **106**, 1142–1148.

44) Sato, B., Spomer, W., Huseby, R.A., and Samuels, L.T. (1979) The testicular estrogen receptor system in two strains of mice differing in susceptibility to estrogen-induced Leydig cell tumors. *Endocrinology* **104**, 822–831.

45) Sato, B., Yamamura, Y., Huseby, R.A., and Matsumoto, K. (1980) Estrogen receptor systems in normal and neoplastic Leydig cells of mice. In "Advances in Sex Hormone Research," eds. Thomas, J.A. and Singhal, R.L., pp. 241–271, Urban & Schwarzenberg, Baltimore-München.

46) Sutherland, R.L. and Murphy, L.C. (1982) Mechanism of oestrogen antagonism by nonsteroidal antiestrogens. *Mol. Cell. Endocrinol.* **25**, 5–23.

47) Terenious, L. (1970) Two modes of interaction between oestrogen and antiestrogen. *Acta Endocrinol.* **64**, 47–58.

48) Weichman, B.W. and Notides, A.C. (1977) Estradiol-binding kinetics of the activated and nonactivated estrogen receptor. *J. Biol. Chem.* **252**, 8856–8862.

49) Welshons, W.V., Lieberman, M.E., and Gorski, J. (1984) Nuclear localization of unoccupied oestrogen receptors. *Nature* **307**, 747–749.

SUBJECT INDEX